Revised and Expanded 2nd Edition

Trans-Oceanic

The Royalty of Radios

John H. Bryant, FAIA
Harold N. Cones, Ph.D.

4880 Lower Valley Road, Atglen, Pennsylvania 19310

DEDICATION

Dedicated to our dads,
Glenn H. Bryant and Harold Cones Sr.,
two professional electrical engineers
who nurtured every aspect of our lives
including our interest in radio.

Designed by "Sue"
Type set in NewBskvll BT

ISBN: 978-0-7643-2838-1
Printed in China

PHOTO AND ADVERTISING CREDITS

All photographs attributed as "Courtesy of Zenith" are reproduced through the courtesy and with the permission of the Zenith Electronics Corporation of Glenview, Illinois, USA. All print advertising published in this work is reproduced here through the courtesy and with the permission of the Zenith Electronics Corporation, as well. All unattributed photographs were taken by and are copyright property of John H. Bryant, AIA. "Zenith," "Trans-Oceanic," and the famous Zenith Lightning Bolt are registered trademarks of Zenith Electronics Corporation.

CAUTION:

Restoration projects both physical and electrical as described herein may be hazardous and are to be undertaken only at the restorer's risk; the authors and the publisher will not be responsible for any injuries or losses arising during restoration projects. Tool usage and electricity can be dangerous.

Schiffer Books are available at special discounts for bulk purchases for sales promotions or premiums. Special editions, including personalized covers, corporate imprints, and excerpts can be created in large quantities for special needs. For more information contact the publisher:

Published by Schiffer Publishing Ltd.
4880 Lower Valley Road
Atglen, PA 19310
Phone: (610) 593-1777; Fax: (610) 593-2002
E-mail: Info@schifferbooks.com

Please visit our web site catalog at www.schifferbooks.com

We are always looking for people to write books on new and related subjects. If you have an idea for a book, please contact us at the above address.

This book may be purchased from the publisher.
Include $5.00 for shipping.
Please try your bookstore first.
You may write for a free catalog.

In Europe, Schiffer books are distributed by:
Bushwood Books
6 Marksbury Ave.
Kew Gardens
Surrey TW9 4JF
England
Phone: 44 (0)208 392-8585
Fax: 44 (0)208 392-9876
E-mail: Info@bushwoodbooks.co.uk

Website: www.bushwoodbooks.co.uk
Free postage in the UK. Europe: air mail at cost.
Try your bookstore first.

TABLE OF CONTENTS

ACKNOWLEDGEMENTS

Any major research and writing effort depends on the cooperation of many individuals if it is to be successful. We consider ourselves fortunate indeed to have had the opportunity to meet and work with so many people interested in our project. At the risk of forgetting someone, we would like to express our thanks:

To Maynard Berger, Carole Dyslin, Howard Fuog, Aimee Huntsha, Eugene Kinney, Howard Lorenzen, Dolores Rutzen, Robert Stender, and especially John I. Taylor, all past or present employees of Zenith, for their valuable assistance in gathering material and information;

To the many Zenith employees we met, who expressed interest in the project and kept us entertained with "Commander" stories;

To Dana Mox, Victoria Matranga, Anthony J. Cascarano, Gordon Guth, Rick Althans, Arthur Pulos and John Okolowicz, who assisted in the development of the industrial design aspects of the Trans-Oceanic story;

To Margaret Budlong Wootton, Richard C. Budlong, Merrily Budlong Holladay, Patricia Budlong Arnold and Dorothy Budlong Skillings, who introduced us to the professional life of Robert Davol Budlong and assisted us as we gathered information about him;

To Jerry Berg, David Clark, Bart Lee, Fritz Mellberg and Bill Wade, five good friends and radio men, who read what we thought was the final manuscript and made valuable comments that helped us shape this finished product;

To Dr. Albert E. Millar, Jr., who proofed the final manuscript for grammar;

To Luther Hall, William Bowers, Harry Helms, Elton Byington, Craig Seigenthaler, Dr. Ronald Moomaw, James Shidler, Gerry Dexter, W. George Chamberlain, Terri Schmidt and Mr. and Mrs. A.E. Peterman, who assisted us and the project at critical times in this two-year project;

To Christopher Newport University, for providing support for portions of the project;

To Sherry Turner, who cheerfully turned all of John's scrawl into manuscript;

To Louise Toole, who located some of our sources, typed from dictation, corrected drafts, proofread, taught us grammar and stayed interested--and most importantly, managed to sing and remain cheerful throughout it all;

To the two Lindas, who understand our needs to undertake such projects, support our efforts, and still admit to their friends that they are married to us;

And, to all our friends and colleagues who have offered words of encouragement.

Our most heartfelt thanks to these friends. If there are mistakes in the manuscript, they are ours, not theirs--as for us, we blame mistakes on each other.

FOREWORD

For forty years, the Zenith "Trans-Oceanic" was a benchmark for the American electronics manufacturing and design industry. In the Trans-Oceanic, science, technology and consumer acceptance came together to provide a product that so impacted society that the name is still recognized by radio enthusiasts everywhere and by most of the American public. This is as it should be, for the Trans-Oceanic was a fabled series of portable radios. Through much of its existence, the Trans-Oceanic was also the most expensive radio offered for sale to the general public. In spite of its high price, the Trans-Oceanic is believed to have been the single most popular radio marque produced during the first fifty years of radio. Many people now involved in the various radio hobbies heard their first shortwave broadcast on a Zenith Trans-Oceanic.

Introduction to the Second Edition

Wow, who could have guessed it? We tried in vain in 1993 and 1994 to locate a publisher for a book dealing with a single kind of radio, the fabled Zenith Trans-Oceanic. One publisher told us that they had to be sure enough copies could be sold before they would agree to publish any book, and even though they thought our manuscript had great potential, they just didn't think they could sell that many copies--ever. Schiffer Publishing, Ltd. decided to take a chance on us, and *The Zenith Trans-Oceanic, The Royalty of Radios*, was released in mid-1995. Now, thirteen years later, it is time to create a second edition.

To us, in addition to documenting the birth and growth of an extraordinary radio, the most important result of the accolades that came from the Trans-Oceanic book was our decision to continue our writing collaboration, and to continue to focus on Zenith Radio. We took advantage of the immense amount of material we had discovered in the Commander's files at Zenith to document the history of Zenith Radio Corporation itself, releasing *Zenith Radio, The Early Years, 1919-1935*, in 1997. Because of the loss of historical archives over the years at Zenith, the details of the earliest years of the company would have disappeared from history had we not acted when we did. Many of the early Zenith pioneers we interviewed in the 1990s have since died, taking the everyday stories of the company with them. And even the fate of the Commander's files are now in question. With the take over of Zenith by LGE, we have lost track of the files and are now working from the tens of thousands of copy pages we made of file material over the years. In 2003, again with Schiffer Publishing, and with the addition of radio collector and historian Martin Blankinship, we released *Zenith Radio, The Glory Years, 1936-1945: History and Products* and *Zenith Radio, The Glory Years, 1936-1945: Illustrated Catalog and Database*. These three books together chronicle the rise of Zenith Radio Corporation from a kitchen table at Chicago's 1316 Carmen Avenue in 1919 to one of the most successful radio manufacturers in America.

Over the years we learned a great deal about Zenith's founder, Eugene F. McDonald, Jr., and his many contributions, not only to the radio business but also to the culture of America (for example, we discovered that when the U.S. Government stopped funding Gutzon Borglum while carving Mt. Rushmore, McDonald supported the effort out of his own pocket.) As we have learned more about McDonald, we began a major effort to make his many contributions known in a variety of formats. We have spoken in many professional venues, including the Naval History Symposium in Annapolis, the Society for the History of Technology, and the American Popular Culture Association. We have addressed countless local, regional and national radio clubs, been on radio and television, and published over 40 scholarly and popular press articles about Zenith and McDonald, including the entries in the *Encyclopedia of Radio*. We have helped a dozen historians worldwide develop projects and we feel we are slowly making inroads on restoring McDonald to his proper place in history. McDonald did not maintain a PR staff with the full-time job of writing about him, as David Sarnoff did. By providing correct information about McDonald at all levels, we feel we will eventually make his many positive contributions common knowledge. Additionally, a series of negative articles voiced by a disgruntled employee and the publication of a negative chronicle of the development of Zenith Radio Corporation in the late 1980s has presented an incorrect--but persistent--view of McDonald that we are slowly correcting.

Among the many corporate materials in the McDonald Files, we found nearly a hundred folders of detailed correspondence about the 1925 MacMillan Arctic Expedition. This expedition, for a variety of reasons, was little known, even among Arctic historians. This surprised us, since it was the first to use shortwave radio in the Arctic, the first to use heavier than air craft in the Arctic, and was Richard Byrd's first exposure to the Arctic. We augmented these file materials with material from the National Archives, the Byrd Archives, and interviews and materials from participants to write, *Dangerous Crossings, The First Modern Polar Expedition, 1925*, published in 2000 by the Naval Institute Press. The idea for the expedition and its use of aircraft was McDonald's, and it was McDonald who acquired the aircraft, facts virtually unknown until the publication of our book.

So, we have been busy telling the Zenith story and it was the success of *The Zenith Trans-Oceanic, The Royalty of Radios*, that has led us to one of our life's passions. For that, we thank you. We, too, remain passionate about the Trans-Oceanic and for some time we have looked forward to sharing the additional information that we have learned in the past decade about "Commander McDonald's personal radio."

Why is all the second edition material at the back of the book?

Even though we see informal and largely undocumented Trans-Oceanic "history," especially at the several Trans-Oceanic web sites, we have found relatively little new documentable history of the Trans-Oceanic that was not in the first edition of this book. In fact, when the publisher asked us to work on a second edition, we researched carefully the file materials we had and decided that the historical portion of the book is largely complete as it stands. There are a few small details that could be added--the first luggage case for the 7G605 was made by Wilt, for example—but the expense of inserting even such a simple statement in the book at the proper place has prevented us from doing so. Adding new material to the back of the book is the most cost-effective means of producing a true second edition that is both affordable and usable to our readers.

Professor John H. Bryant, FAIA
Stillwater, Oklahoma

Distinguished Professor
Harold N. Cones, Ph.D.
Newport News, Virginia

ZENITH
TRADE MARK

Radio Companion

*A six-tube set completely self-contained!
No need to open case to operate
No exterior loop or antenna required*

Listening, with the keenest pleasure, to music and voices in the cities they have left behind!

Lively orchestras entertain these boys, miles and miles from civilization.

Receiving the latest market reports, the latest news developments, with the aid of the Zenith Radio Companion.

Zenith—MacMillan's Choice
Encased in a Light Traveling Bag!

Here's a six-tube radio set that's entirely self-contained — tubes, "A" batteries, "B" batteries, loud speaker and loop antenna complete, and it's a *Zenith!*

Packed into a small, beautifully finished traveling case — much smaller than the average suitcase —this new Zenith is the most compact set ever made giving clarity, quality, volume and distance.

Do you see those two little buttons close to the handle? Those are the controls. In order to operate the new Zenith Radio Companion you simply turn the controls to bring in the station you want—then for maximum volume you swing the case so that the loop is facing that particular station. You will be astonished at the clearness with which the music and the voices come through—and in what volume!

Think what it would mean to you to be able to take one of these new Zeniths with you on your travels and outings. A real radio set—the exclusive choice of Donald B. MacMillan for his Arctic expedition — yet so compact that it takes up no more space than a light traveling bag!

Think of the fun you could have with this set—the dance music you could listen to on moonlit nights—the orchestras that would play for you as you and your pals gathered round the camp-fire—the companionship it could give you on your motor parties—at the bathing beach. Picture the enjoyment it could bring your guests at the house-party or the week-end gathering.

Again, think how such a set would while away a lonesome evening in that dreary out-of-town hotel—what a godsend it would be to that invalid mother —to that dear relative or friend who must spend weeks and months in the hospital!

You will want to know more of this remarkable set—so light and compact, so easy to operate, so wonderfully convenient. No earphones, you understand. No outside antenna. Yet clarity, volume, quality, distance! A real Zenith, packed into a traveling case!

The coupon will bring you full particulars.

Zenith Radio Corporation
**McCormick Building
Chicago, Illinois**

The height of luxury —motoring to music!

When three is company at the bathing beach.

A constant source of entertainment and delight to invalids.

Chapter 1
THE PORTABLE RADIO

The popularity of the Zenith Trans-Oceanic radio was due in no small part to its portability. It represented a logical evolutionary step in the development of portable radios which had begun in early 1923 when Edward Armstrong, the inventor of much of modern radio technology, used a portable radio to demonstrate a new circuit to RCA engineers.[1] Armstrong, it is said, strolled off the elevator in the RCA building carrying a self-contained portable radio which was playing an opera program. In the case of Armstrong's receiver, the use of the word "portable" was somewhat questionable. His set was 18" x 10" x 10" and was very heavy--but it was self-contained and was capable of receiving radio signals while being carried.[2]

Major E. Howard Armstrong and his wife Marion listening to what may have been the world's first portable radio. This radio was constructed by Armstrong and presented to his wife as a wedding present. c. 1923.

There has been a great deal of debate as to whether the advancement of science drives technology or technology drives the advancement of science. In the development of the portable radio, the line in this debate is fuzzy since the whole science of radio developed over a very short time period. "Wireless" leapt into science after the work of Maxwell in the 1860s, Hertz in the 1880s and Tesla and Marconi in the 1890s and early 1900s. Fessenden's Christmas Eve program of 1906 marked the earliest broadcast of a human voice to an audience and in 1920, a scant 14 years later, KDKA became the first station to provide regular broadcasts. By 1923, there were 500 stations! The interplay between the development of science and the development of this technology was thus a close one.

Like all mass-produced products, the development of the portable radio was not only dependent on science and technology, it was driven by consumer acceptance. Consumer acceptance is derived from a variety of sources, not the least of which are cultural and social forces. During the early days of radio, massive cultural changes were occurring: women gained more freedom; the internal combustion engine drove the development of the automobile, encouraging society to become mobile; humans took to the air; there was a world war; and, the industrial revolution was in full swing, providing everyone with a wide variety of manufactured products to fill needs many did not even know they had until advertisements told them so.

The development of the portable radio was also guided by a cast of "characters" unlike those associated with any other product. Among them were the eccentric genius Nicola Tesla, tower-climbing Edwin Armstrong, science fiction/radio whiz Hugo Gernsback, and flamboyant salesman and explorer Eugene McDonald.

The emerging profession of industrial design was then added to the mix of science, technology, social changes and unique personalities. Industrial designers turned the very utilitarian appearance of early radio into beautiful design, increasing its desirability to the consumer and thus its place of prominence for the future. The evolution of radio was a spectacle of events and personalities that came together in a very short period to produce a medium that quickly became a standard of everyday life.

By mid-1923, RCA was producing the $97.50 Radiola 2, regarded by some as the first portable radio. It was not totally self-contained, however, since it required an external antenna. Also in mid-1923, a small independent company produced a quasi-portable radio called the Operadio 2, which sold for $190. It was a somewhat unique radio in that the front door could be removed and attached to the top of the receiver to form a loop antenna, thus eliminating the need for an external antenna. The Operadio 2 could not be played and carried at the same time, however, and it could not be carried with the loop antenna attached to the top. Although a self-contained unit, the Operadio 2 is not viewed as a true portable radio.[3]

Zenith Radio Corporation produced a number of experimental portables in the very early days of the company. One of these experimental units was tested by Commander Donald B. MacMillan in the Arctic in 1923 and then introduced to the public in 1924 as the Zenith *Companion*. The Companion is generally regarded as the first totally portable radio[4] since it was self-contained and could play with the briefcase-type cover closed.[5] At nearly the same time (or perhaps a month earlier), a virtually identical radio was produced by the Westburr Company of New York. It will probably never be known which of these radios was the first truly portable radio, but by August 1925, there were 22 portables of one fashion or another on the market. With the improvement of radio circuitry and programming, the conventional radios (consoles and table models) became much more popular and the novelty of the poor-performing portable wore off. Surprisingly, by 1927, the market for portables had almost totally dried up.[6]

Interior view of the Zenith *Companion*. The *Companion* was considered the first truly portable radio since tuning and volume adjustments could be made without opening the top, it could be carried while it played and it did not need an external antenna.*Courtesy of Zenith.*

It was technically possible to build a quality portable in the early twenties, however. In 1922, the National Bureau of Standards developed a self-contained 6-tube portable radio for research in radio direction finding. The circuitry and the sound quality were good enough that, had it been marketed, it could have been a successful radio.[7]

Only a few portable models were produced between 1927 and the late 1930s. Many of these radios were more properly called "universals," since they worked on either electricity or storage batteries, much the same as the farm radios of the day. During the 1930s, automobiles became more popular and affordable within a broader segment of society and "outings" became more frequent. It would seem logical that this new mobility would be an important stimulation in the development of the portable, but the radios typically taken on these outings were the better performing standard radios that were set up on running boards or on tables with wires run to the car battery. The portable never did gain a large following during this period.

One would think that the novelty of a portable radio would lead to its mass acceptance in the early days of radio, but such was not the case. Although possessing a handle, the early portables were difficult to carry; they were large and heavy and had poor sound quality and sensitivity. Almost as important, the radio programming of the day was not interesting enough to encourage the average person to carry a heavy portable around in order to hear it. Early radio did not affect as many lives as it would in the future; in later years, listening to the radio, and particular programs on the radio, became daily events, but this was not generally so in the early 1920s.

The Depression ended the 1920s radio fad, but people began to listen more to the radios they had. By the late 1930s, as the world changed and the prospect of another world war loomed, the country became obsessed with the news and any small radio that could be easily carried on outings or in an automobile was a desirable radio. As a result of this new demand, a number of companies began producing truly self-contained, battery-powered portable radios. The first of these appeared in mid-1938 and represented the vanguard of what was to follow. By the end of 1939, 150 portable radio models were available to the consumer.[8]

Commander Donald B. MacMillan with the Zenith *Companion*, touted as the world's first truly portable radio. MacMillan tested the *Companion* during his 1923 Arctic expedition and Zenith introduced it to the public in 1924. *Courtesy of Zenith.*

Commander Eugene F. McDonald, Jr. admiring the precursor of the Trans-Oceanic, the 1940 Model 5G401 and the newly released Model H500 Trans-Oceanic, c.1952. Photo taken at the mantel in The Commander's office at Zenith headquarters. *Courtesy of Zenith.*

The rapid proliferation of portables was made possible by several new developments. In 1939, electronically efficient 1.5-volt filament tubes were developed which greatly extended battery life while providing quality comparable to the larger home radios. The low battery drain also made portables economically practical to a wider audience. Loktal base tubes were introduced about the same time. The lock-in design meant that tubes would not jar loose as portables were being moved about, thus greatly increasing the dependability of the radio.

Among these early portables was the medium wave Zenith model 5G401, a luggage-styled radio with a fold-down door that bore a striking resemblance to the all-wave portable that would be introduced by Zenith two years later. The President of the Zenith Radio Corporation, Commander Eugene McDonald, Jr., wanted to have a radio that would bring him world news no matter where he traveled on his many adventures. Zenith engineers began development of a super-portable to meet his needs in the fall of 1939, starting with a model 5G401 "right off the line."[9] More than two years and over twenty working prototypes later, they produced the 1942 7G605 Trans-Ocean[ic] Clipper, the world's first all-band truly portable consumer radio. Although the Clipper was only produced for three and one-half months before the assembly lines were shut down for war production on April 22, 1942, it was wildly successful and became the first of a long line of Zenith Trans-Oceanic portables.

Screen star Myrna Loy and the Zenith model 5G401 portable radio for 1940. The chassis of the 5G401 formed the basis for the early shortwave portable prototypes which led to the Trans-Oceanic Clipper. *Courtesy of Zenith.*

As the transistor era dawned in the 1950s, radio manufacturers soon saw the solid-state technology as being ideally suited for portables. The world's first transistorized set, a shirt-pocket three transistor portable by Regency, was more a novelty than a radio. It was, however, a clear demonstration of the potential of transistorized portable radios. One of the most successful of the early portables was the shirt-pocket Zenith Royal 500, manufactured from 1955-59.

A Zenith engineering team spent two years in intensive development before the Commander personally approved the design for the first transistorized Trans-Oceanic. Zenith introduced the Royal 1000 Trans-Oceanic in December 1957. It is important to note that Zenith continued to make tube-type Trans-Oceanics in parallel with the Royal 1000. Three essentially identical tube-type Trans-Oceanic models were produced during the early transistor years: models Y600 (1956-1957), A600 (1958) and B600 (1959-1962). The B600 Trans-Oceanic was the last portable tube radio manufactured in America.[10]

The twenty years between 1957 and 1977 are generally thought of as the "Transistor Years" in portable radios. By the late 1970s, individual transistors were being rapidly replaced by integrated circuits which combined many electronic components on a single chip. These twenty years were also the time of intense international competition in the American portable radio market. At the end of the tube era in 1962, a foreign-manufactured portable radio was a novelty in the U.S. Exactly eighteen years later, Zenith was the last remaining manufacturer of American-built portable radios. When Zenith moved production of the model R-7000-1 Trans-Oceanic from Chicago to Taiwan, the American manufacture of portable consumer radios came to a close.[11]

The history of the Trans-Oceanic is a history, in microcosm, of the post-war American radio industry. During most of those years, the Trans-Oceanic was the most expensive consumer radio available in the U.S. It varied in price from $600 to well over $1400 (when expressed in 1994 dollars). Price notwithstanding, the Trans-Oceanic is also believed to have been the single most popular American radio marque.

The passing of the Trans-Oceanic brought to a close an era begun so long ago by Howard Armstrong's portable demonstration in 1923. Throughout the life of the Trans-Oceanic, few people realized either its historical importance or the fact that this very special radio was a product of the lifestyle and interests of Commander Eugene F. McDonald, Jr., himself.

During World War II, all American radio manufacturers shifted their design and production capacity to the war effort; few, if any, consumer radios were manufactured between 1942 and early 1946. By 1946, the pent-up demand for consumer products was enormous. This was especially true for portable radios, which quickly came to represent the new affluent and very mobile American lifestyle. All major radio manufacturers responded by developing portable radios covering a wide range of designs and prices. These new portables were heavily advertised with glamorous advertisements associating the radio with picnics, convertibles, trips to the beach and pretty girls: the portable radio was a resounding success!

The demand for Zenith's Trans-Oceanic was particularly strong. During the war, the exploits of the Trans-Oceanic Clipper and its reputation as an almost indestructible radio had become widely known. In the post-war era, the Trans-Oceanic became the darling of the international "jet set," and the treasured companion of world travelers and armchair adventurers. Most Trans-Oceanic advertising was positioned in the upscale *Holiday* magazine and in *National Geographic*.

Until Commander Eugene McDonald's death in 1958, the Trans-Oceanic continued to be "his" radio. He was personally involved in the design and the marketing of each model, supervising Zenith engineers and independent industrial design consultants as they developed subsequent Trans-Oceanic models.

Even though both RCA and the shortwave radio-oriented Hallicrafters Company produced competing models from time to time, the Zenith Trans-Oceanic completely dominated its niche in the market as *the* American all-band radio. Its performance was so superior that most people interested in listening to radio broadcasts from abroad purchased a Trans-Oceanic and simply used it as a table radio. The purchase of Trans-Oceanics by strictly armchair adventurers was such a significant market that it is very likely that the majority of tube-type Trans-Oceanics were never used as portables at all.

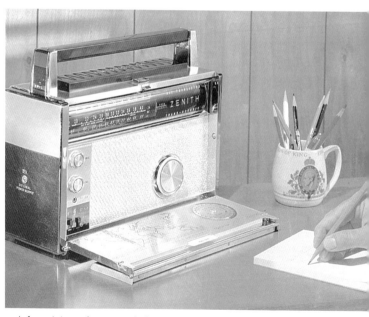

Advertising photograph for the Royal 3000-1 Trans-Oceanic. Photo c. Fall 1963. *Courtesy of Zenith.*

Opposite page photo:
Advertising photograph for the R-7000 Trans-Oceanic. Photo c. Spring 1979. *Courtesy of Zenith.*

RADIONICS

with its subdivisions of Electronics, Radio, etc., says "Look into the record of Portable Supremacy."

THIS IS THE FIFTH OF A SERIES of radio merchandising questions on post war planning.

It covers the subject of PORTABLE RADIOS.

Portable radios have grown to be a big volume factor in every aggressive radio dealer's selling picture.

Zenith recognized this fact EARLY. Zenith has consistently LED THE INDUSTRY in percentage of its famous Wave-magnet portable sales—to the industry's portable radio sales. Indisputable figures prove that 'fact.

Zenith created a sensation when it brought out its Transoceanic Clipper at $75.00 retail.

Immediate and amazing acceptance proved to the country's radio merchants that portable business need not mean small unit of sale.

Portable value never before given . . . portable performance never before engineered . . . and portable unit price never before dreamed of—combined to achieve the greatest world-wide acceptance ever accorded *any radio receiver.*

That's history—not claims . . . and you know it! It hardly seems necessary to frame the questions. Nevertheless, in your post war planning, please be sure to add this check-up to those which Zenith has been giving you in previous advertisements of this series. If you haven't seen them, we shall be glad to mail them to you on request.

Questions:

1. What radio manufacturer has the greatest reputation for extreme value, amazing performance and reliability in the portable field?

   ☐ Write name of corporation.

2. What is the best known individual portable model in the country?

   ☐ Write brand and name of model.

3. What is the highest priced nationally sold quality portable on the market?

   ☐ Write brand and name of model.

We hear you saying: "Why ask when we know?" It is good for you to see your *own answers* in black and white. Put them down!

They should fix indelibly in your mind the phrase "Zenith Has Portable Supremacy." Figure that after the war Zenith Portables *must be* "SUPREMACY PROTECTED!"—by astonishing innovations—

—and you will have the answer to YOUR POST WAR PORTABLE POTENTIAL.

ZENITH RADIO CORPORATION, CHICAGO 39, ILLINOIS

BETTER THAN CASH
U. S. War Savings Stamps and Bonds

ZENITH
•LONG DISTANCE• **RADIO**
RADIONIC PRODUCTS EXCLUSIVELY—
WORLD'S LEADING MANUFACTURER

Chapter 2
THE COMMANDER AND THE TRANS-OCEANIC
How a Lifestyle Led to the Development of the World's Most Popular Portable Radio

The first Zenith Trans-Oceanic portable radio was built specifically for Commander Eugene F. McDonald, Jr., President of Zenith Radio Corporation. The development of the go-anywhere Trans-Oceanic was a logical evolutionary step in the adventurous lifestyle of Commander McDonald, a larger than life figure in the history of radio. His legendary quests for adventure and invention at times make it difficult to decipher truth from fantasy. However, his contributions to "radionics" (his own term for radio electronics[1]), are numerous and well-documented. His achievements were acknowledged posthumously by his peers on April 4, 1967, when his name was entered in the Broadcast Pioneers Hall of Fame. Among the accomplishments listed in the citation were his role as Founder, President and first Board Chairman of Zenith Radio Corporation, his dynamic merchandising strategies, his inventions and innovations, his role as explorer and his role as the first President of The National Association of Broadcasters. He was also cited for having established one of the nation's first radio stations, WJAZ, and for pioneering the development of shortwave radio, international communication, ship-to-shore, FM, VHF and UHF television, radar and subscription television.[2]

Commander McDonald's rise to fame is a true Horatio Alger story. He was born in 1886 in Syracuse, New York[3], his father variously remembered as either an insurance salesman[4], a storekeeper[5] or a railroad worker.[6] At the age of 12, the younger McDonald received his first exposure to electricity as a doorbell installer's helper. He was not interested in high school, leaving after two years to pursue work. Before he left, however, he organized a grievance strike that resulted in a demonstration and a street parade over some long-forgotten subject.[7] In 1904, he went to the Franklin Auto Company in Syracuse (the company rented a building from his father) where he worked for a week as a mechanic, filing metal castings eleven hours a day. From there he moved to the Engine Building Department where he worked up to assembling an entire engine in a single day; then, because of his flamboyance, he was transferred to the Sales Division.[8]

He quickly became known for his promotional activities and his ability to obtain press coverage. He arranged, for example, to drive William Jennings Bryan around the city of Atlanta in a Franklin and to appear in most of the photographs.[9] While driving a Franklin automobile up Lookout Mountain, Tennessee, on another of his promotional events, he fractured his skull in an accident which left him permanently deaf in one ear.[10] He also went on the road opening Franklin dealerships and became the Southern Sales Manager for the company. He left Franklin after five years and joined the Imperial Motor Company of Buffalo as Sales Manager selling Packards, Pope-Hartfords, Buicks and Babcock Electric cars.[11]

In late 1910, McDonald moved to Chicago to handle the business dealings of a good friend, Charles Hannah, who had invented a self-starting device for automobiles which operated by exploding acetylene in the cylinders of the engine. Although the self-starter was an idea whose time had come, the company went broke and took with it all of the money that McDonald had accumulated in his earlier endeavors.[12]

McDonald turned his talents at that point to selling used cars. He approached the Detroit Electric dealer in Chicago with an incentive scheme which would help them reduce their large inventory of used automobiles as well as make money for himself. He obtained used cars from the dealer, repaired them mechanically, painted them, and sold them for a profit. It was a successful partnership yielding McDonald $16,000 (and the dealer $5,000) in a year, a large amount of money in 1911. In order to promote the venture, he turned again to his flair for publicity. He drove one of his new used cars up the steps of the Logan monument in Chicago's Grant Park, arranged for a photographer to be present and paid a policeman $10 to arrest him.[13]

In 1912, McDonald made automotive history when he organized a finance company to sell Ford automobiles on credit.[14] Many items were sold on credit at that time, but not automobiles. Only the wealthy owned automobiles, the prevailing feeling being that if you could not pay cash, you should not own one. McDonald's scheme was considered foolish for a period of time by his contemporaries. However, he was soon selling over 20,000 automobiles a year, making at least $50 on each car, plus ten percent interest. His method was simple: he bought automobiles new, sold them on time, discounted the paper at the banks, and amassed the profits.[15] This endeavor formed the foundation of the McDonald fortune and was the first of many McDonald "firsts."

At the outbreak of World War I, the 31-year old McDonald sold his time-payment business and joined the Navy.[16] While in Chicago, he had purchased a "telegraphone," a wire-recorder device that could be used to record telephone conversation without the knowledge of the caller. By 1917, the U.S. manufacturer had gone out of business. The Navy was interested in the device and when they found that McDonald understood and could operate it, they offered him an officer's commission. He assumed the rank of Lieutenant-Commander and worked with the Naval Intelligence Corps investigating sabotage cases.[17] McDonald left the Navy in 1919 and briefly managed real-estate properties while looking for a new business venture. His discovery of that venture is one of the most repeated stories in radio.

The Beginnings of Zenith Radio Corporation

On New Year's Eve 1920, Commander McDonald went to a service garage of a man named Layton in Chicago's south side to pick up his Packard prior to a holiday trip to Syracuse to visit his mother. He noticed a small group of people listening to what he thought was a phonograph. When he wondered aloud why so many people would be standing around on New Year's Eve listening to a phonograph, he was told that it was a new device called a "radio."[18] He listened to the broadcast of KDKA in Pittsburgh and learned that the radio, an early model Grebe about the size of a shoebox, sold for $200.[19] When he found out that these "radios" were difficult to acquire, and sold for a high price, he decided to go into the radio business.[20]

Investigating, he discovered that to manufacture radios it was necessary to hold a license from E.H. Armstrong, the inventor of the regenerative circuit. He also discovered that Armstrong had temporally suspended issuing the licenses. McDonald's plan to sell radios seemed stymied.[21] Further investigation during 1921 led him to the north side of Chicago, to two young men, Karl Hassel and Ralph Mathews, who were operating a small radio manufacturing business from a kitchen table in one of their mother's kitchens.[22] The two had been experimenting for some years and had reached the point in radio manufacturing where they could construct one set a day. The set sold for $200.[23] The crucial part of the equation, as far as Commander McDonald was concerned, was that they held the essential Armstrong patent. Hassel and Mathews' radio building business exercised their license under the name "Chicago Radio Laboratory." In addition to manufacturing radio sets, Hassel and Mathews also operated an amateur radio station, Station 9ZN, and called their radio set the "Z-Nith" after their amateur call.[24] Hassel and Mathews had more orders than they could possibly fill for the Z-Nith and McDonald offered to become a financial partner in their undertaking. A partnership was formed, with McDonald becoming the General Manager.[25]

Commander Eugene F. McDonald, Jr. preparing to read the news to Commander MacMillan's 1923 Arctic Expedition over the shortwave facilities of Zenith station WJAZ. *Courtesy of Zenith.*

As an aid to marketing, McDonald changed the hyphenated name of the set to Zenith and, in 1923, the partnership was incorporated as the Zenith Radio Corporation. Several years later, in July 1927, Zenith had grown to the point that they were awarded the first license ever issued by the Radio Corporation of America. This license allowed them to manufacture sets using all patents held by RCA, as well as those held by the American Telephone and Telegraph Company, Westinghouse Electric and Manufacturing Company, and General Electric Company.[26] The Chicago Radio Lab, organized in 1915, thus became the Zenith Radio Corporation in 1923, with Commander McDonald as President of the corporation, and within a few years it had developed into one of the largest radio manufacturers in the world.[27] [Commander McDonald was proud of his accomplishments and a number of documents exist in his private files that detail this early history in his own words.[28]]

Since the Armstrong patent was licensed to the Chicago Radio Laboratory, it became the manufacturing arm and Zenith Radio Corporation became the marketing arm for the radios. It was not until several years later that the two merged so that both manufacturing and marketing could be carried out by Zenith Radio Corporation. On February 20, 1923, Carl E. Hassel filed the official trademark of the Zenith Radio Corporation, the famous Zenith Lightning Bolt, with the U.S. Patent Office.[29]

Promoting The Business And The Man

Ever the merchandiser, in 1923, the flamboyant Commander McDonald met Commander Donald B. MacMillan, a well-known arctic explorer, and persuaded him to put a Zenith shortwave radio on his arctic exploration schooner *Bowdoin.*[30] The project was very successful for both the explorer and Zenith. In 1923, McDonald built one of the nation's earliest radio stations, station WJAZ, at the Edgewater Beach Hotel manufacturing site in Chicago. During the next MacMillan expedition, McDonald read the world news and letters from relatives and friends to the Arctic exploration team every Thursday night over the shortwave

facilities of WJAZ.[31] For extra impact, McDonald arranged for MacMillan's sister, Lettie[32], to relay a message to the expedition. McDonald himself even accompanied the expedition to middle Labrador before returning to broadcast from Chicago. The public was totally caught up in the adventure and large audiences listened every Thursday night. This was not only the first on-site news broadcast, but was a powerful selling tool for shortwave radio in general and Zenith shortwave radios in particular. For Arctic explorer MacMillan, the 1923 Expedition was his first expedition which was able to stay in constant touch with civilization.

Another McDonald "first" occurred in 1923 when he was instrumental in founding the now-powerful National Association of Broadcasters. The fledgling organization consisted of only five stations and was formed to augment program resources and to defeat the legal suits over royalties of the American Society of Composers, Authors and Publishers by building a shared music library.[33] In the early days of the organization, McDonald encountered the editor of a radio fan magazine who was complaining about surplus issues remaining when a planned magazine merger had not occurred. McDonald offered to mention the magazine on all five of the stations in the broadcasters' association for a $1,000 contribution to the NAB. The scheme worked and within 24 hours the magazine was sold out. This is considered to be the first case of "network" commercial sales and earned Commander McDonald the dubious distinction of being the pioneer of paid network radio advertising.[34]

McDonald, always looking for a way to sell his products by expanding the market, found the restrictive radio laws of the mid-20s not only a deterrent to Zenith but also to the advancement of the medium of radio. Herbert Hoover, the Secretary of Commerce, held what amounted to a one-man dictatorship over the air waves, deciding what stations would be licensed, their broadcast schedule and their location. Hoover's power was derived from the inability of the Federal government to pass a radio control law, and from a majority of broadcasters who felt that some control was better than none. Zenith was allowed to broadcast only one hour a week, from eleven to twelve and had to share the frequency with KOA in Denver.[35]

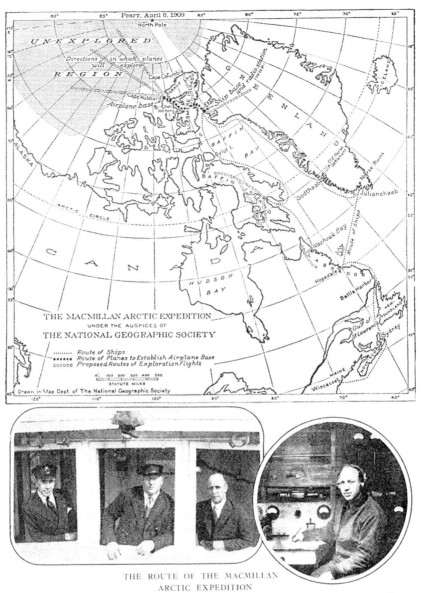

THE ROUTE OF THE MACMILLAN
ARCTIC EXPEDITION

At present in Arctic waters. The map shows the route of the two ships, the *Peary* and the *Bowdoin*. The cut at the upper left shows the command of the expedition in the pilot house of the *Peary*. At the left is Lieut. Commander E. F. McDonald, Jr. (U. S. N. R. F.), in command of the *Peary*, next is Captain George Steele of the *Peary*, and at the right is Commander Donald Macmillan, in charge of the expedition. The insert at the right shows John L. Reinartz, radio operator in his cabin aboard the *Bowdoin*. Radio communication with the 40- and 20-meter transmitters aboard both ships has been established with amateur operators in the United States, Canada, and England from the Greenland base. Short waves are used because they are less subject to attenuation in daylight. The expedition during its entire time in northern latitudes will be in constant daylight. 2 GY, the short wave station maintained by RADIO BROADCAST has been in communication with the *Peary*, WAP, using a wavelength of 40 meters at her Etah, Greenland base

Information on the 1925 Arctic Expedition as published in the
October 1925 *Radio Broadcast* magazine. *Courtesy of Zenith.*

McDonald felt strongly that the assignment of times and frequencies was discriminatory. The grounds for a historic legal test case were set when he protested to the Commerce Department in 1925 demanding to be able to broadcast on one of six channels reserved by agreement for Canadian broadcasters. Permission was not granted, but Zenith began broadcasting on a Canadian wavelength anyway, forcing the U.S. Government to bring action.[36] On April 16, 1926, the case was decided in United States Circuit Court in McDonald's favor, proving that the radio frequency allocation laws of the time were unenforceable.[37] McDonald then appeared before a Senate Committee[38] and suggested the formation of a Federal Radio Commission (later to become the Federal Communications Commission.)[39] The actions of Commander McDonald thus were at least partially responsible for the proliferation of radio stations, 24-hour program availability and the watchdog federal agency which was established to oversee the expansion.[40]

The Navy of the mid-1920s conducted a number of "show of force" cruises to all parts of the world and McDonald saw the opportunity to attempt to convince the military of the superiority of shortwave equipment over the distance-limiting longwave equipment then in use. He persuaded Admiral Ridley McClain, the Director of Naval Communications, to commission Fred Schnell, a radio amateur recommended by the American Radio Relay League (and a friend of Commander McDonald's), for assignment to the cruiser *U.S.S. Seattle*, then the flagship of the Pacific Fleet. The Pacific Fleet was planning a major cruise at that time and McDonald felt that proving the usefulness of shortwave radio communications to this Pacific-wide cruise would be beneficial not only to his own radio sales but also to the Navy in general. The Navy, told that some radio experimentation would occur, was not aware at the time that they had been set up for one of McDonald's most dazzling promotional activities.[41]

Commander McDonald broadcasting traditional Eskimo songs to the U.S.S. Seattle off Tasmania during the 1925 MacMillan Arctic Expedition. *Courtesy of Zenith.*

Commander McDonald's previous work with explorer MacMillan led to his being invited to be second-in-command on MacMillan's 1925 National Geographic Arctic Research Expedition; McDonald was put in command of the *Peary*. He equipped the *Peary*, as well as the other MacMillan ships, with Zenith shortwave equipment. Three U.S. Navy airplanes under the command of Richard E. Byrd were attached to the expedition and were also equipped with Zenith shortwave equipment.[42] The superiority of shortwave equipment was demonstrated early in the expedition when the *Peary* stopped at Godhaven on Disco Island, a Danish possession, to refuel. McDonald was told by the dockmaster that fueling could not commence until authority was received from a Danish official in Washington. That permission would be difficult to obtain since the longwave transmitter in Washington could not send that far in daylight, and arctic night would not fall for three months! McDonald responded by turning on his shortwave apparatus and contacting an amateur radio operator in Washington, who went to the Danish Embassy, got the permission, and transmitted the permission back to Disco Island in less than four hours.[43]

The expedition then proceeded as far north as Etah, Greenland, where McDonald unleashed the plot that he had planned earlier. He sent a shortwave message to the now-Lieutenant Fred Schnell (the amateur radio operator he had situated with the Pacific fleet) on the *Seattle*, then located on the coast of Tasmania, over 12,000 miles away. With his usual flair for the dramatic, McDonald had Eskimos sing into the microphone. He was told by Morse code from the *Seattle* that the music sounded like college yells, to which McDonald responded that it did sound like college yells and the signal must be getting through fine.[44] This transmission, from 11 degrees south of the North Pole, to the deep Southern Hemisphere near Tasmania, had tremendous impact on the Navy, who then began purchasing Zenith shortwave equipment. For many years thereafter, Zenith products were advertised as "the choice of arctic explorers" and Admiral MacMillan was still being featured in Zenith advertising 30 years later.

The arctic expedition with MacMillan had an interesting sidelight. Wherever McDonald went during the expedition, he gave away short-wave radios, with extra batteries, to nearly everyone he met. When he returned to Chicago, he received numerous letters from Greenlanders asking for more batteries or an alternative method of powering the radios they had received.[45] McDonald had thought often of the problem of battery depletion since there were also many radios on unelectrified farms in the U.S. that would profit from some form of battery extension. As McDonald himself stated in "The Arctic Inspires a New Market," an undated manuscript from the Zenith archives,[46] "Statistics showed me that there were more than 11 million unelectrified homes in the country. What a market!" His idea was to use the wind, and when he heard that two young men in Iowa, John and Gerhard Albers, had perfected a wind-powered battery charger, he ordered one for testing. The device consisted of a small windmill that drove a rewired automobile generator and thus could be used to charge conventional automobile batteries, which could then be used to power radios.[47] McDonald went to Iowa to meet the inventors and purchased 51 percent of their stock. He then ordered 50,000 "wind chargers" and by perfecting mass production techniques, reduced the price of the wind charger from $40 to $15. The Winchargers, a common sight on the Great Plains for the next 40 years, were often sold for $10 with the purchase of a Zenith radio.[48] The two young men continued to work with the Wincharger Corporation, a subsidiary of Zenith Radio Corporation, for some time, providing their product to eighteen major radio manufacturers worldwide.[49]

In 1931, McDonald initiated TV research at Zenith and established the first all-electronic TV station, W9XZN, in 1939. By 1941, Zenith was broadcasting with TV. In 1940, Zenith also pioneered one of the world's first FM stations, WEFM.[50] Commander McDonald stayed very much involved with both the business and creative end of Zenith operations and personally held a number of patents for unique electronic devices.[51]

Adventures And Adventuring

Commander McDonald's non-radio adventures are as interesting as his radio adventures and often resulted in new products or techniques

Dealer card advertising special pricing for the Wincharger when purchased with a Zenith 6-volt farm radio. *Courtesy of Zenith.*

for Zenith. For instance, on a gold hunting expedition to Cocos Island, he and an engineer perfected a device that used radio waves to detect metal. The humidity of the island, however, saturated the coils of the device, rendering it useless to detect gold. McDonald perfected a method of protecting the coils from the humidity so that the device would work on future expeditions, and thus developed humidity-proof coils.[52] His resultant pioneering work in humidity-proofing various electrical instruments was credited with making a major contribution to the high degree of reliability of American radio products during the Second World War.

In addition to his much publicized Arctic exploration, Commander McDonald organized several scientific and exploring expeditions and frequently returned from these trips with plant and animal specimens for zoos and museums.[53] On one such trip to the South Seas on his 185-foot yacht, the *Mizpah*, the party "discovered" Dr. Frederick Ritter and Dore Strauch on the Galapagos Island of Floreana. The two had come to Floreana in search of Utopia but had fallen on hard times when McDonald found them. The Commander left supplies and rendered assistance before continuing his adventure.[54]

Commander McDonald also used the *Mizpah* as a test platform for a variety of radio gear.[55] A radio-related expedition on the *Mizpah* to Labrador investigated the effects of the *aurora borealis* on radio propagation. Later, the Commander produced and published a chart of McGreagor Bay by using then-new electronic depth finders and ranging instruments.

Among Commander McDonald's other "firsts" in the electronics industry were his production of the Radio Nurse, a monitoring device that was placed near an infant's crib with the receiving speakers at various locations in the house[56]; a Great Lakes telephone service; the first farmer's radio, which was a radio designed to work off a single auto storage battery; the first single tuning control for radios, which took the place of three or four individual controls; the first totally electronic radio that ran on house current; the first radio sets to have more than one speaker; the first pushbutton radio; and, of course, the Wavemagnet loop antenna.

Development Of The Trans-Oceanic

With McDonald's adventuring background and radio pioneering knowledge, his desire for a "high powered" portable radio to accompany him on his far-flung adventures was almost inevitable. On June 27, 1939, McDonald received patent number 2,164,251 for the Wavemagnet, an external loop antenna that allowed radio reception in trains, airplanes and metal enclosed buildings.[57] This antenna was attached to a window glass by suction cups and connected to the radio by means of a cable. Although the Wavemagnet greatly enhanced signal strength under some circumstances, the fact that Zenith and all other existing portable radios tuned only the standard AM broadcast band still prevented good reception in remote locations. This limitation was particularly troublesome to McDonald on trips to his Canadian fishing lodge on upper Lake Huron.

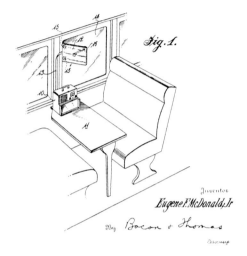

May 14, 1940.   E. F. McDONALD, JR   2,200,674
                          RADIO APPARATUS
                        Filed June 26, 1939

Fig.1.

Inventor
Eugene F. McDonald, Jr
By Bacon & Thomas
Attorneys

Commander McDonald's patent drawing for the Wavemagnet antenna used with a portable radio. *Courtesy of Zenith.*

Copy from a 1942 dealer's catalogue showing a full-color counter display featuring the Wavemagnet as used with the Universal Model 6G601 portable. *Courtesy of Zenith.*

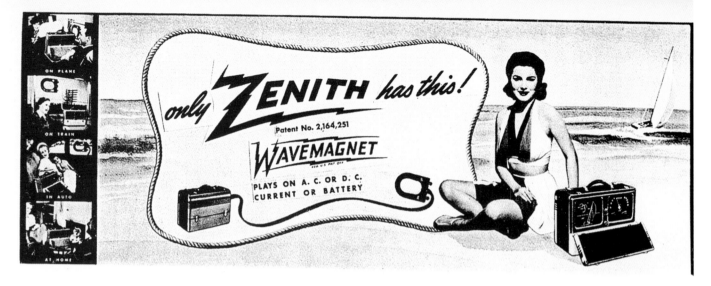

Copy from a 1942 dealer's catalogue showing a full-color
"window streamer" display featuring the Wavemagnet.
Courtesy of Zenith.

McDonald's yacht *Mizpah* was one of the largest on the Great Lakes and traveled for extended periods of time on the Lakes, often to remote locations. Reception on the Zenith portable that McDonald carried was almost nonexistent in some of the areas he frequented, and he became interested in finding a portable that would work. Not only did he want to hear the radio news from Europe, where Hitler was assuming power, but he wanted to keep abreast of the weather. Since his Zenith broadcast band portable did not prove powerful enough for his needs, he asked the Zenith staff to develop the ideal travel radio. In a November 9, 1951, memo, Commander McDonald recalled the final incident that led to the development of the industry's first shortwave portable simply: "Standard broadcast reception at my Canadian fishing lodge was extremely poor...."[58]

The Commander's response to this poor reception, although alluded to many times in his files, is best detailed in a personal Christmas letter to friend Herman Staebler dated December 18, 1942:

I ordered our laboratory to make this short-wave portable on August 2, 1939. I sent a radiogram from my yacht when in McGregor Bay telling our chief engineer exactly what I wanted. The laboratory built about twenty different models and submitted them to me, all of which I rejected until I received two of the final laboratory samples a year ago last June. I sent one of these sets North with Comdr. Donald B. MacMillan into the Arctic. I took the other one to Canada with me. I received a radiogram from MacMillan saying that in his twenty years in the Arctic with radio he'd never been able to keep in touch with the outside world as he had with this little portable. I had the same success with it in Canada. I then authorized them to go into production. We were not ready for deliveries until January 1, 1942--over two and one-half years of laboratory work is behind this job that I'm sending you.[59]

The Zenith design team responsible for the first shortwave portable included engineers Gustafson, Passow, Striker and Emde.[61] Howard O. Lorenzen was a development engineer at Zenith in 1939 and worked on the first prototype with this team. In a November 24, 1992, interview with the authors, Mr. Lorenzen recalled the incidents that led to the first Zenith Trans-Oceanic:

I went down to production and took one of the regular portables that they had in production and I just tore the broadcast band out and put in shortwave coils. When I did that I designed everything for the shortwave band that had the stations we wanted...For the shortwave antenna I made a one-turn loop the size of the cabinet and coupled it with capacitors. It worked pretty well and I was surprised that we could pick up the BBC and other shortwave stations.

When I got it so that it worked fairly decent, Gustafson [Gilbert E. Gustafson, Chief Engineer at Zenith Radio] and I took it up to the Commander in his office. The Commander said, "I will give this a good trial because I am getting ready to make another run up to [Canada]."

When he came back, he called Gus and I up to his office. We sat around a big table and the Commander said, "I was up there and you know, that damned thing works." I said, "Well, I thought it would." He said, "I could hear the Voice of America [sic], BBC in London, and I could even hear Berlin stations and got Nazi news direct from Germany"... the next thing he said was, "We're going to go into production with this and we are going to call it the Super Deluxe Trans-Oceanic Shortwave Portable." I said, "Remember, Commander, that is strictly a shortwave portable." "Oh," he said..."it will also have a broadcast band."

He then pulled a leather covered case out from under his desk and said, "I got in touch with one of the suppliers to give me a leather covered case to put this thing in." He handed the case to me and I said, "This doesn't fit any of our existing chassis." "Well, you can make it whatever you want," he said. I took it and the rest of the time I worked with Zenith I tried to get a shortwave and broadcast band in the case together. But those volt-and-a-half [powered radio] tubes were horrible. They didn't have any gain and if you put a band switch in, everything just kind of quit working. After the first one I sweat for several months before I left Zenith trying to solve those problems. [Interview question: Do you know how many prototypes were made of the receiver before it went into production?] No, I don't, because I left Zenith for the Naval Research Laboratory in 1940. But they must have had a lot of trouble getting all the bugs out of the thing.[61]

Commander McDonald not only field tested the Trans-Oceanic prototype in Canada, he also tested it in early October 1941 on the Baltimore and Ohio's Capitol Limited and on Eastern Airlines planes on Washington to Miami and Miami to Chicago flights. He reported that over southern states standard radios could not bring in the World Series reports, but that passengers were treated to the scores through a shortwave link of WGEA, Schenectady. He also provided reception from London, Berlin, Rome and Moscow.[62]

Opposite page top:

A very early "sailboat" version of the 1942 Trans-Oceanic Clipper. *Courtesy of Zenith.*

Opposite page bottom:

The "bomber" model of the Trans-Oceanic Clipper. *Courtesy of Zenith.*

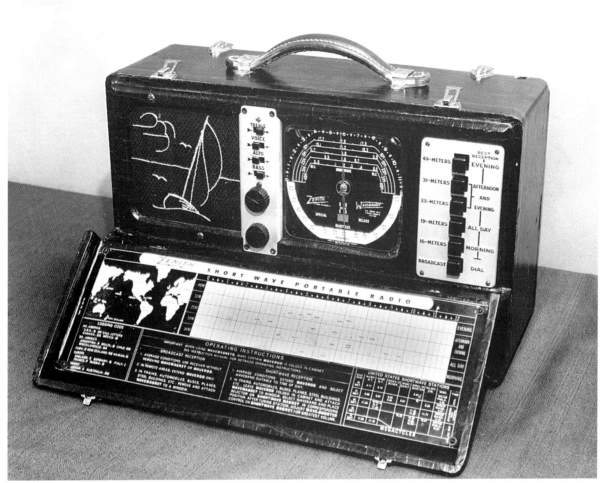

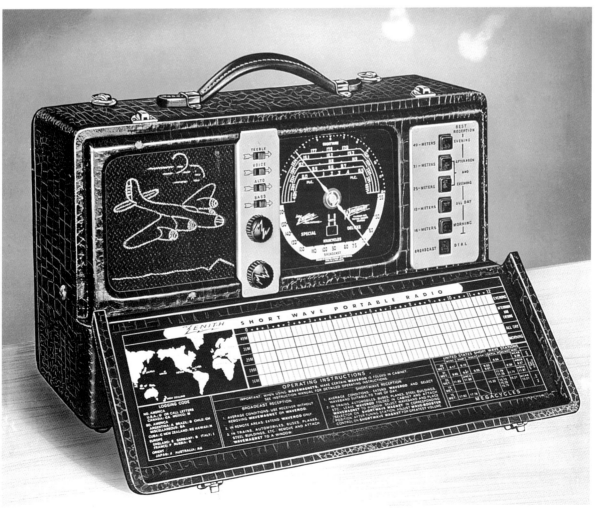

"THE TRANS-OCEANIC CLIPPER TAKES CHICAGO BY STORM. Window display at the Hudson-Ross department store featuring the Zenith Trans-Oceanic Clipper Deluxe." Verbatim caption from February 1942 Zenith press release. *Courtesy of Zenith.*

The Zenith Trans-Oceanic Clipper, containing McDonald's additional patent (2,200,674), the *shortwave* Wavemagnet, a tuneable antenna, went into limited production in October 1941[63] and began deliveries to the public in early January 1942. A Zenith distributors release dated December 17, 1941, stated:

> On September 17th we wrote you a letter telling you of our development of a brand new super deluxe portable.

> At that time, although our distributors responded enthusiastically to the prospect of having a portable "with everything on it" to sell, we were still undecided as to just when we should add this new model to the line, since our manufacturing division was already taxed to the limit under the most unusual conditions which prevailed.

> Then-on December 7th-came the Japanese attack on Pearl Harbor. People rushed to equip themselves with radio. Portables took an amazing upward sales spurt.

> In view of this condition, and since the purchasing division has through the autumn months been successful in gathering together the necessary parts-we have decided to put this new number into immediate production.

> A few, not many, will come off the lines between now and Christmas. These will be sent to distributors as samples. Following Christmas and the first of the year regular production will lead off with the new portable and carry along with it the regular Universal Portable, too...we are calling the new portable-the ZENITH TRANS-OCEAN [sic] CLIPPER...[64]

> ...We have priced the deluxe portable "with everything on it" at $75 list price. We had hoped to achieve a lower list, but with the many features we have built into this set-that has proved impossible.[65]

In late December, just before Christmas, Commander McDonald sent "a copy of my latest 'baby'" to many of his business and personal friends along with a letter detailing the history of its development. He received enthusiastic responses from the recipients.[66]

The new radio was introduced in a media splash in early January 1942. In Chicago, the radio was introduced by filling a window of the Zenith downtown showroom at 680 North Michigan with a Trans-Oceanic display. A Miss Kelsey, the originator of the display, rendered assistance to retailers in town and soon Hudson Ross had a duplicate window at their Randolph Street store. Marshall Field and Saks Fifth Avenue also installed a Trans-Oceanic window.[67] Pictures were taken of these windows and sent to dealers for promotional purposes and the window display pieces were offered for sale to distributors.[68] By January 24, 1942, 30,000 Trans-Oceanic portables had been released.[69]

In early January, a "bug" was discovered in the Trans-Oceanic which caused Commander McDonald to write to all those to whom he had sent a complimentary radio. The text of a typical letter went:

> Glad you like my new "baby," the short-wave portable radio. However, I have been using it down in Florida and down in the Keys for the past two weeks. I have found a "bug" in it. If you take the set into these extremely humid countries, the shortwave end loses its sensitivity after a few days exposure to extreme humidity.

> If you happen to take yours South and experience a loss of sensitivity, don't go out and buy new batteries for it. Just open up the back and put it in the hot sun with the back open for about an hour or two. Its complete sensitivity will return.

> If you don't have the sunlight, put it in the kitchen or the galley and let the set get thoroughly dry...This is a "bug" I didn't discover until I played with it in the South. This will be corrected in the next run.[70]

The Commander's approach to researching this problem was to have his opened Clipper put in his humid closet in Florida for four days, then wrapped in oil cloth and flown to Chicago for tests.[71]

Production of the Clipper continued until April 22, 1942, when the factory, under War Planning Board Federal edict, converted totally to war production.[72] At the time of conversion, 35,000 Clippers had been produced and there were 100,000 unfilled orders for the shortwave portable; those that had been produced were in service worldwide.[73] Marketing surveys suggested that the majority had been sold to people in the "above average income group."[74]

The development of the very first Zenith Trans-Oceanic portable was thus the reflection of the lifestyle of a flamboyant adventurer who needed a powerful shortwave portable to accompany him on his explorations. The radio developed by the Zenith staff to meet Commander McDonald's needs became an often imitated industry standard over the years and enjoyed the longest continuous run of any radio model. It continues to enjoy a huge following today.

"LAST ZENITH PORTABLE SEES RADIO INDUSTRY 100% FOR WAR -- Commander E.F. McDonald Jr., founder-president of Zenith Radio Corporation, accepts the last civilian radio, "for the duration," from factory manager W.E. Fullerton and cheering Zenith workers in Zenith's big plant. From now on Zenith goes full time on radio war work. Zenith has been rapidly stepping up war production for months past." Verbatim caption from April 1942 Zenith press release. The ceremony was held on April 22, 1942. *Courtesy of Zenith.*

A thousand letters
WHICH BREAK OUR HEARTS

LETTERS!
—THE PILE IS GROWING—

—letters from Privates, Corporals, Colonels, Generals, Seamen and Admirals—from Wacs, Waves, Spars . . . from everybody . . . everywhere.

Their urgent pleas strike a universal note . . . they say in effect:

I know of only one portable radio that will do the work out here—they say, writing from Africa and Alaska . . . from Australia and the South Seas . . . from all over the globe—Only one . . . and that is your Zenith Transoceanic Short Wave Portable Clipper. My folks tell me they have tried everywhere to obtain one with no success. Can you help me?

. . . so these letters come to us.

To each request must go the answer "No"—an unwilling "No"—and our regrets that this must be so. We were over 100,000 sets oversold on this one model when we ceased civilian production for 100% war work.

Nothing would please us better than to have a great plenty of these justly famous portables to ship to all who need them—especially at this Christmas season—when our thoughts turn to loved ones everywhere. For our Transoceanic Portable Clipper is a real friend to the men and women in the service. Those who managed to get them early feel themselves fortunate; they are the envy of their friends!

BUT THESE PORTABLES MUST WAIT. The entire Zenith organization is now engaged in giving all its efforts to the making of tremendous quantities of urgently needed radionics* materiel for the armed forces. These things must come first—even ahead of the tender link with home which a personal radio provides for the fighting man far from friends and family. Thus we help to speed the day of "absolute Victory"—help to bring closer the next real American Christmas . . . with "Peace on Earth—Good Will to Men" . . . when families shall be reunited—and when home life can once again resume the even tenor of its ways.

Chapter 3
THE TRANS-OCEANIC
IN WAR AND PEACE

The Trans-Oceanic Clipper was produced just in time to be purchased by soldiers, sailors and airmen and to accompany them to the far-flung battles of World War II. At home, the Trans-Oceanic served to connect the home front to the war and served as a vital blackout radio.[1] Almost from the day of its introduction in 1941, the Zenith Trans-Oceanic Clipper amassed a fiercely loyal following. Owners spoke highly of the Clipper and wrote often and freely to the Zenith factory to tell of the exploits of their radios.

The Trans-Oceanic in The Second World War

World War II proved to be a valuable and rugged testing ground for the new radio. Photos and testimonials came to Zenith from all parts of the world during the war years; many of them were reprinted in the Zenith Radio Corporation newspaper, the *Radiorgan*. The March, 1945 *Radiorgan*, for example, had the following letter:

...When the racket of falling earth and debris was over, down in our foxholes, we heard the strains of "Star Dust" gradually reassert themselves on the beach at Lunga. We peered out and saw that the tent where we had been sitting was gone. Personal gear was scattered far and wide. But in the midst of the ruins that Zenith portable radio had never missed a note.....It continued 'giving out' when others had quit.

Another testimonial in the same issue of the *Radiorgan* read:

This set was first taken to Africa in the invasion at Oran in November, 1942. It was dropped in the Mediterranean and was taken across Africa in a truck from which it was dropped while the truck was going at a high rate of speed. The odd shaped bulge at one end of the set is a makeshift 'A' battery since we were unable to get the proper replacement.

This set returned to combat on June 6, 1944, when it was brought to France early on 'D' Day.

In addition to the soaking and rough handling, it has been necessary to rewire three tubes in the set as there were no replacements and it was necessary to use G.I. tubes. I might add that the wave magnet was lost months ago and we had to make an antenna coil of our own. Then when it was dropped in the ocean it was necessary to replace the volume control with a replacement from a German tank radio. Part of the R.F. coils were rewound from an Italian radio part. In spite of all the rough treatment my Clipper has had, it still plays well.

Thousands of similar reports came to Zenith during the war years attesting to the almost unbelievable ruggedness and versatility of the Clipper. Stories of the Trans-Oceanic continuing to work after receiving brutal abuse were proudly related. One Clipper was blown into a water-filled shell hole at Guadalcanal, dried in a cook's oven, and put back in service; another was badly charred but continued to function. One Clipper had this adventure:

...Needless to say, we feared the worst for our radio, you know, salt air, moisture and worst of all, the fact that those B bags were piled like sacks of wheat in the hold. Well, when we arrived in the beautiful Hawaiian Islands, we dug out that Zenith, turned it on, and, WOW, there came one of those wistful island tunes from a Honolulu station, and so it went, from is-

land to island, always in the B bag in the hold. Then there were the times, many of them that the fellows in the tent would get to scuffling and, BOOM would go our Zenith on the floor, be it wood or earth. At first, we feared to pick it up, at least it would be broken, but twist the knob, and there would come station WXLI Saipan and good old Saipan Sam. Well, it finally happened, somehow, somewhere, we lost the hinge door on the front, but we fashioned one from the end of an orange crate. Had to have a door on it for those B bag trips, you know. Now, I could go on and on about that Zenith portable, but to conclude with, I would just like to say that after some thirty-eight months overseas, it didn't look much like the one we left the States with, but the thing played on and on...[2]

After the war, Zenith used a number of testimonials in their Trans-Oceanic advertising campaigns to enhance the image of their radio as a rugged and faithful companion. One such ad contained the following testimonial from a British soldier:

...During the whole war, it [the Trans-Oceanic] was either with me or close behind with my luggage.

From then 'til now, the set has accompanied me to Kenya, Tanganyika, Ethiopia, British, Italian and French Somalilands, Sudan, Egypt, Greece, Palestine, Aden, India, Burma, Malaysia and last but not least, England.

Besides being under enemy fire, it has stood in the open in blistering heat and sandstorms, likewise in monsoon, rain and snow. It has been carried on camels and mules and trucks and armored cars, in trains and ships. I have watched it dropped into cargo-holds by coolies, and have sympathized with it after suffering a kick by a donkey. On at least one occasion in the bush, a lion has had a good sniff at it.

As a result of, or perhaps in spite of, such treatment, there has been no occasion during its ten years life that the set has failed to work and for 95% of the time has been absolutely first class. Even today, people comment on its tone and performance.

A 1945 Zenith ad titled, "Back Soon, Better Than Ever: The Portable Radio Sensation The War Stopped," announced that production of the Trans-Oceanic would resume as soon as parts became available after the war. The ad proudly stated, "G.I.'s write pleading letters by the thousands from overseas. They wanted a Zenith Trans-Oceanic Clipper to hear home broadcasts like other 'G.I.'s' who had pre-war Zenith Clippers. It broke our hearts to refuse them... but war production came first."

When production of the Clipper had to be stopped in April 1942, to allow Zenith to enter the war effort full-time, about a thousand of the coveted Trans-Oceanics were kept for special presentations.[3] A 1944 *Radiorgan* issue shows a smiling Frank Sinatra receiving a Trans-Oceanic Clipper. The radios were unavailable to the general public during the war and because of the testimonials of the servicemen, were highly coveted. Even as early as 1944, the mystique of the Zenith Trans-Oceanic was beginning to grow.

After the war, testimonials took the form of yachtsmen and sea captains praising the Trans-Oceanic. For example, B.A. Jacobson, Master of the *S/S Constitution*, stated in a Zenith ad:

I thought you would like to know how much I enjoy my Zenith Trans-Oceanic radio. It certainly makes for pleasant relaxation.

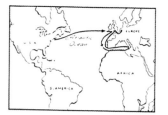

ZENITHS FIGHT ON ALL FRONTS

Vets of Two Theaters Take It . . . Ask for More

Here are the almost unbelievable actual war experiences
of two Zenith Clipper radios as retold by their owners
who took them to battle on opposite sides of the globe

THE CLIPPER owned by Lt. (j.g.) Cameron W. Babbitt was a tropical traveler. If ever a climate ate the heart out of a man and the guts out of a radio, it's the dark, humid tropics. But the Zenith Clipper remained the one link with home for thousands of sailors and marines in this isolated wilderness. But let the lieutenant tell you in his own words:

. . . "When the racket of falling earth and debris was over, down in our foxholes we heard the strains of "Star Dust" gradually reassert themselves on the beach at Lunga. We peered out and saw that the tent where we had been sitting was gone. Personal gear was scattered far and wide. But in the midst of the ruins that Zenith portable radio had never missed a note.

"It seemed, at times, that not only was this radio of superb construction but that it bore a charmed life. An example: A sneak air raid caught us listening to 'Tokyo Rose' (or her predecessor) making extravagant claims of our annihilation between excellent recordings. We scattered to our foxholes, neglecting to turn the radio off. A bomb had landed within twenty feet of the tent in which we had been listening to that beloved song. After that, this radio became even more of a personality to us.

"This Zenith radio was purchased new in Oakland, California in July, 1942. It was used, within the space of 10 months, in widely separated places, under the most adverse conditions conceivable and still continued to function as if new. In all, it has accompanied me over fifty thousand miles, providing enjoyment all the way.

"Its first scene of operation after it was acquired was in New Caledonia where it was the center of entertainment for thousands of soldiers, sailors and marines, in those early days when our armed forces were moving into Caledonia. This was to be repeated continually in the future.

"After a few weeks in Caledonia we moved (the radio and I) to Espiritu Santo where, under the coconut trees and surrounded by jungle growth, it was again the center of entertainment for as

This One Was Bombed In the Pacific

A VOICE FOR "TOKYO ROSE"
Lt. C. W. Babbitt, SC, USN, purchased his Zenith Shortwave Clipper in Oakland in July, 1942. Since then it has carried the voices of "home" to countless sailors and marines in far-away Pacific isles. It also brought them the voice of "Tokyo Rose," whose treacherous monologue was so exaggerated that it provided entertainment. Despite travel, climate and bombings, the guts of this Clipper asked for more.

many as could get within sound of its speaker. Within a week I left for Guadalcanal, leaving the radio behind as I thought the conditions of the future were so uncertain that it should be where it could give maximum enjoyment to the maximum personnel.

"However, upon arrival at Guadalcanal it was found that the need for just such entertainment outweighed all other considerations. Accordingly, a bomber crew returning south was requested to bring the radio up on their next trip and in due time it arrived. (The crew reported great success in using it on the trip up.) The radio was hooked up to a line supplied by a captured Japanese generator.

"In spite of the fluctuation of power supply, (the batteries were long ago consumed) the frequent jarring of bombings and bombardments, the intense, damp heat, and the continuous use it received, the radio never faltered. In the early days (September '42) it was our only source of amusement, and in the later days was still the center of our recreational life.

"The incidents, tragic, amusing, discouraging and heartening, in which this radio figured are too numerous to mention. The comfort it brought to thousands of fighting men who gathered around it in those months is beyond words. It, and the all too infrequent letters, brought to us the pictures and remembrances of

This One Fell Into the Mediterranean

WAR-SCARRED VETERAN
. . . of the African campaign, Mediterranean and Atlantic immersions, the "D" Day Invasion of France is shown with its scars and a makeshift "A" battery. It has been rewired for GI tubes; its present volume control is from a German tank. An Italian radio contributed new coils. The wavemagnet, which was lost, is strictly GI. Owner T/5 Jacob Lockwood says it still plays, receiving programs direct from the States.

home and our loved ones that are so needed to carry on in the face of the uncertainties of war.

"It was not the only radio on Guadalcanal, but thousands thought it was the best. It continued 'giving out' when others had quit.

"If the Zenith Radio Corporation continues to manufacture products of such outstanding quality—and I am sure it will—there are many new friends—friends made by this one product of the skilled, efficient hands of the employees of that corporation, that will be clamoring for your products. You who conceived, planned, built and delivered this radio have the gratitude of thousands for the pleasure it gave us."

JUST ABOUT everything bad that can happen to a radio did to that owned by T/5 Jacob F. Lockwood. But it has kept going strong and was, when last we heard, bringing our home broadcasts direct to Belgium. Lockwood felt so good about it that he wrote the following "History of a Clipper."

"This set was first taken to Africa in the invasion at Oran in November, 1942. It was dropped in the Mediterranean and was taken across Africa in a truck from which it was dropped while the truck was going at a high rate of speed. The odd shaped bulge on one end of the set is a makeshift 'A' battery since we were unable to get the proper replacement.

"This set returned to combat on June 6, 1944 when it was brought to France early on 'D' day.

"In addition to the soaking and rough handling, it has been necessary to rewire three tubes in the set as there were no replacements and it was necessary to use G.I. tubes. I might add that the wavemagnet antenna was lost months ago and we had to make an antenna coil of our own. Then, when it was dropped in the ocean, it was necessary to replace the volume control with a replacement from a German tank radio. Part of the R.F. coils were rewound from Italian radio parts. In spite of all the rough treatment my Clipper has had, it still plays well."

CLIPPER SERVES ON JAP ISLE

Official U. S. Navy Photo
U. S. Postoffice on Kwajalein was complete with up-to-the-minute reports via the Zenith Clipper shortwave portable. Clippers have served on morale duty around the globe, wherever U. S. men put down their duffle. Thousands of letters attest to their effectiveness and dependability under conditions like this.

Back Soon - Better than Ever

the Portable Radio Sensation the War Stopped

1.

Remember the Transoceanic Clipper? Only Zenith had it and it really excited America! Here at last, was a portable radio with "big set" performance. Folks flocked to buy it. But then came the war . . . and production stopped.

2.

"G. I's" wrote pleading letters by the thousands from overseas. They wanted a Zenith Transoceanic Clipper to hear home broadcasts like other "G. I's" who had prewar Zenith Clippers. It broke our hearts to refuse them . . . but war production came first.

3.

War production and how! Compact Radionic power is our specialty. So, our job was to develop secret Radionic devices for planes, tanks, ships where space is at a premium. In the process we discovered new ways to concentrate enormous power in tiny Radionic units. You'll call it a touch of genius.

4.

Now the "Clipper's" coming back . . . better than ever. Only Zenith will have it. More powerful, streamlined and beautiful. It will be completely re-designed for easier carrying . . . around the house . . . on outings. Its rich, full tone will match big consoles. And its performance will make the whole world your neighbors!

5.

Will work where others won't because of new super-powerful Zenith Wave-magnets. You'll get local stations clearly and powerfully at home, on trains, in planes, aboard ship and even at remote vacation spots. And short wave programs from this country and abroad with power and distance that will amaze you.

Naturally Zenith is "Tops" in Portables Because Zenith Concentrates On

RADIONICS EXCLUSIVELY

Zenith's leadership in portable radios is the direct result of 30 years of concentrating on Radionics Exclusively . . . the science of radio waves. No one but Zenith can make that assertion. Out of this vast reservoir of specialized knowledge and skill will come portable Radionic creations, so superior that you instantly recognize the touch of genius.

Watch for the portable radio sensation the war stopped. It's coming back better than ever—and only your Zenith dealer will have it.

BUY VICTORY BONDS - BETTER THAN CASH

For the Best in Radio Keep Your Eye on Zenith

Zenith ⚡LONG DISTANCE⚡ **RADIO**
REG. U.S. PAT. OFF.

RADIO • FM • RADAR • TELEVISION • RADIO-PHONOGRAPHS • HEARING AIDS

THE CHOICE OF "THE VOICE"

Fresh from the vault where it was stored for just such an occasion, a Zenith shortwave Clipper portable radio is presented to "The Voice," Frank Sinatra. The donor is Stuart Loucheim, treasurer and general manager of the Motor Parts Company, Philadelphia distributors of Zenith radios, who also do a big business with Columbia records including the Sinatra classics. The admiring audience is made up of retail sales people who deal in Mr. Loucheim's and the Swoon King's Columbia records.

One thing that particularly interests me is the weather band. I am surprised and delighted to find such a useful feature in a portable radio.

We of American Export Lines say that the *Constitution* and her sister ship, the *Independence*, bring modern American living to ocean travel and I find that the Zenith Trans-Oceanic radio is a fitting adjunct to modern American living.[4]

"Dear Zenith...It was worth more than all the gold in Korea..."

The Korean conflict touched off another flurry of letter-writing by Trans-Oceanic owners, primarily soldiers in the field. The Korean testimonials took on much the same tone as the Second World War testimonials. For example:

...I spend [sic] a year, 1951-52, during the original days of the Peace Talks with H Co. 3rd Battalion, First Marine Division, as a bazooka gunner. In your picture, there is such a fellow on the left-hand side. [This is a reference to an advertising cut that appeared in various magazines that bore the caption "Peace or War - a Group of GI's of the famed 27th Infantry Regiment, one of the first to arrive in Korea, gather around a Zenith Trans-Oceanic standard broadcast and shortwave portable radio for the latest reports on the Peace Talks. Their future may well depend on the outcome of the truce negotiations now under way in Kaesong].

Our Section - seven men - of bazooka, bought a Zenith Trans-Oceanic radio from the PX on one of our few Reserve periods back behind the lines. The radio was carried whenever we went on the front lines and while back in Reserve. That radio was the bright spot in our life over there, and the boys from our company were constantly kept up to date on the news in Korea and the world.

We could pick up news broadcasts from Japan and know what the rest of the front lines were doing, and also we heard much good music from Japan which was a wonderful moral [sic] builder.

So, ever since I have been home from Korea, my choice to my friends and myself has been a Zenith. Someday I hope to own another wonderful Zenith Trans-Oceanic.[5]

A more unusual "testimonial" came from the Iron Triangle on the 18th of July, 1951:

We have a complaint. Yesterday at about two-thirty P.M. 250 pounds of T.N.T. went off in our area and blew our Trans-Oceanic Radio into many pieces. After all it was at least 15 feet away. Seriously, gentlemen, we are very low today for the preceding did actually take place. As usual our radio was on but in the hasty departure from the immediate area of all concerned the radio was left sitting on our mortar bunker. The 4th platoon had chipped [in] and a member brought it back from Japan. For three weeks it had been the pride and joy of our whole company. After wearing the battery out in the first few days everyone was out scraping up Army BA 70s, 40s and 39s and were then hooked up in various combinations we managed to keep it running day & night. You can never guess what a joy that radio was. It was worth more than all the gold in Korea. For the last four weeks we had been on Hill 1073 (meters, that is), in a defensive position. We are right on the front line and our patrols make daily contact with the enemy. Life has absolutely nothing to offer anyone in Korea and the only outside contact we had was our mail. Only once in a while do magazines come up on the hill and even more seldom do we get down below and there is nothing there except a good nights sleep. The radio brought life to us that nothing else could. We ate up the news, music and other programs beamed to us from Japan and other parts around the world. I have seen other Trans-Oceanic sets around over here and had experience with them in the States. As you know it is the finest portable made and we thought you would like to know just how far they travel and how much they mean to a few of us who have little else. It might even prove to be good advertising. If you think I am writing in hopes of a replacement you're absolutely correct.[6]

"Korea: When these UN tankers are not engaged in routine patrols, they tune in their radio for the latest development in the current peace negotiations at Panmunjom." United Press press release, 12/29/51. *Courtesy of Zenith.*

Members of the 7th Division Mortar Group listening to the U.S. election results on a Zenith Trans-Oceanic, November 1952. *Courtesy of Zenith.*

One of the most remembered Trans-Oceanic advertisements:
"Astride a camel in Pakistan on November 15, 1954, Captain
Ransom Fullinwider, USN, is shown with his Zenith Trans-
Oceanic model that has circled the globe." *Courtesy of Zenith.*

Zenith saw to it that numerous photographs of the G500 being used by soldiers circulated in the media. For instance, one photo, showed tank soldiers sitting on the turret of a tank listening to a G500. The caption for the photograph, dated December 29, 1951, was, "When these UN tankers are not engaged in routine patrols, they tune in their radio for the latest developments in current peace negotiations at Panmunjom." Another news photo was captioned, "U.S. troops tuned in the Super Trans-Oceanic last June to get latest news of the UN-Communist peace negotiations." And another photo was captioned, "Yesterday: Zenith's Traditions of Service...A World War II veteran, Zenith's Trans-Oceanic multi-band portable radio saw front line service again during the Korean conflict." The Newark Sunday News for November 9, 1952, showed a group of soldiers gathered around a Zenith Trans-Oceanic radio; the caption read, "Members of the 7th Division Mortar Group remain ready for action on Korean front as they get election results on a portable radio."

The importance of the Zenith Trans-Oceanic to the Korean soldier was perhaps best exemplified by its presence on the set of the television series M.A.S.H., where it made occasional appearances throughout most of the years of the series. According to a Zenith press release, the Trans-Oceanic was even featured prominently in a Department of Defense film produced in the Korean War years called "Guardian Angel." The film had 400 showings in Class A time and played in several theaters in New York.[7]

One of the more unusual exploits for the Trans-Oceanic was its use (or near use) in the Bay of Pigs invasion of 1961. Zenith was asked on two-week notice to design the packaging and bagging so that 250 of the military version (model R-520/UUR) could be air dropped into the landing zone for "troop" use. Although it was a major undertaking, Zenith officials were never told if their design was employed or not.[8]

In The Field With The Trans-Oceanic

After the Korean conflict, wartime exploits gave way to geographical and anthropological exploration and Trans-Oceanic testimonials began to come from a large number of expeditions. A Zenith press release showed the radios being enjoyed by tribesmen in Kenya, Ethiopia, Sudan and Pakistan. It was noted in one of these testimonials that a native chieftain in Kenya was particularly fond of listening to Bing Crosby on the Zenith Trans-Oceanic. In another news release, natives in the French Cameroons, West Africa, were shown intently listening to the voices and music brought in by a G500 Zenith Trans-Oceanic. A December 1952 National Geographic article titled, "Bhutan - Land of the Thunder Dragon," featured a photograph of a Zenith Trans-Oceanic (model G500) being admired by the Bhutanese.[9] The American Polar Basin Expedition of January 1955 was among a number of polar expeditions to take the Trans-Oceanic portables with them.

Sir Edmund Hillary, famed Mount Everest explorer, took a Zenith Trans-Oceanic transistorized model Royal 1000 on several of his expeditions, including a Yeti hunting expedition with TV personality and zoo director Marlin Perkins.[10] The 1958 American Karakoram Expedition used the Trans-Oceanic model Royal 1000 and filed the following testimonial:

We are very happy to report that the Zenith Trans-Oceanic transistor receiver which you supplied our expedition worked extremely well under all circumstances. It readily withstood the abuse of a three week march through hot dusty terrain, and glacial cold. Reception was excellent despite the fact that we were surrounded by some of the largest mountains in the world.

We used your receiver daily to obtain important weather reports and it was constantly tuned to music and news. In spite of this, it was never necessary to change batteries in the two months of use.

The light weight made it possible for us to carry the set beyond Base Camp to an altitude of 18,000 feet where its performance remained excellent.

We are entirely satisfied with this radio, and we are happy to give it our full endorsement.[11]

Mount Everest climber Sir Edmund Hillary (c) and zoo director and television personality Marlin Perkins (r) hold the Trans-Oceanic model Royal 1000 that will accompany them on their summer 1960 expedition to the Himalayas to hunt the Yeti. *Courtesy of Zenith.*

A detailed report of the performance of the Zenith Trans-Oceanic transistor model Royal 1000 was filed with this testimonial. The expedition team made several modification suggestions concerning strengthening the antenna hinge and the need for additional support for stability. It is interesting to note that these same refinements were made in the next major redesign of the radio.

The Trans-Oceanics not only held up well to the cold of polar expeditions[12], but also withstood extreme desert heat, as attested to by the Lab Geodetics Corporation in an April 22, 1964, letter to Zenith Radio Corporation:

Lab Geodetics Corporation...specializes in astral observations with a device called Zenith Camera (no relation). Reception of the U.S. Navy time signal WWV is of paramount importance to us and since our work is basically all in foreign areas it is also a major problem.

On a number of polar expeditions where the same requirement time signals existed, we used large expensive communications sets specifically set up for this purpose until some fellow came along with a Trans-Oceanic under his arm at a time when we were having an unusual amount of trouble with reception. The results were three-fold:

1. We received WWV beautifully on the 1000-D and set our chronometers.
2. We gave this big and heavy communications system to the communications section and got 1000-D's.
3. We had no more time problems in the Antarctic mapping operation.

In 1962 we bought three new 1000-D's for our foreign operations. The first task undertaken was eight months in Egypt's Western Desert with two of them, (with temperatures up to

135 degrees F. in the daytime)--no problems. Immediately on completion of the Egypt job, the same two receivers were sent to Honduras where during the course of normal events a marsh buggy was turned over and the two receivers went to the bottom of the river--subsequently fished out, they were used without cleaning or repair for the remainder of the job....

A couple of points are particularly significant about these remarkable gadgets:

a. They are transported all over the world by air, but on a job they travel by jeep, donkey, camel, Arabic dhow, fishing boat, and what not.
b. Since the day they were bought, not a single repair of any kind has been made on either of the two.
c. While both units look like refugees from the scrap heap, they work well and are scheduled for another trip overseas.[13]

Such testimonials received again and again by Zenith Radio Corporation attested not only to the success of the Commander McDonald-inspired receiver but also to the fact that the Trans-Oceanic had become a way of life for its users. People looked upon the Trans-Oceanic as more than just another radio. They viewed it as a hard-working family member and a companion to their adventures, both real-life and armchair. This dedication to the sturdiness and quality of the Zenith Trans-Oceanic, and the emotional attachment of its owners, insured its position as one of the longest continuous line of radios in radio history. A growing number of people are collecting the Trans-Oceanic. For some, it is a return to another time; for others it is the discovery of a world-class receiver that cut its teeth on many of the extraordinary events of the 20th century.

The Trans-Oceanic model G-500 in use in the Sahara. *Courtesy of Zenith.*

Chapter 4
DESIGN AND STYLING OF
THE TRANS-OCEANIC

Prior to the era of mass production, the "look" of household goods was determined by the artisans who handcrafted them. In the early industrial period, the appearance of goods mimicked the historical handcrafted styles. By the mid-1930s, an entirely new profession had developed to service the needs of industry for the design of new mass-produced products--the profession of *industrial design*. The first generation of these professionals sprang from backgrounds in the visual arts (painting, sculpture) or from the ranks of Depression-idled architects. Industrial designers worked closely with engineers and sales staffs to determine market needs and then to design and/or "style" products for mass production by the giants of American and European industry. In the post-war era, these first generation industrial designers were joined by graduates of many new professional education programs in industrial design.

Happily for Zenith, Chicago was one of the main centers of industrial design practice and education in North America.[1] Today, design professionals examining the early Zenith line of the late 1920s and 1930s easily conclude that the appearance of most of these radios was drawn directly from furniture design of the day--large radios were copies of large furniture. Smaller table model sets were heavily influenced by clock and jewelry cabinets. By the mid-1930s, however, at least a few Zenith radios were being designed by professional industrial design consultants; cabinets began to "look like radios" rather than various other pieces of furniture.

Robert Davol Budlong

Several different industrial design consultants may have worked with Zenith in the latter half of the 1930s. By far the most prominent of those in Zenith history was also one of the founders of the industrial design profession itself, Robert Davol Budlong. Budlong, a 1921 graduate of Grinnell College, a highly regarded liberal arts school in Iowa, studied art at two institutes in Chicago and opened an advertising agency in Chicago before turning to industrial design. Budlong transitioned his activities to industrial design around 1933, with commissions from the Hammond Clock Company and other midwestern manufacturers.

In either 1934 or 1935, he was introduced to Zenith through his friendship with one of the Zenith engineers, a Mr. Wagenknethz.[2] Budlong related years later that, at the time, he felt that Zenith radios were technically very good but that their appearance and ergonomics could easily be improved by applying the principles of industrial design. He called on Commander Eugene McDonald five times before the Commander agreed to retain him on a trial basis as an industrial design consultant.[3]

This relationship, begun rather reluctantly by Commander McDonald, led to a lifelong friendship between the two men and to the creation of some of the classics of American product design. The close alliance was thought unusual at the time, for the two men were of radically differing personalities. Commander McDonald was a gruff, hard-driving, demanding captain of industry while Budlong was a quiet, self-effacing "gentleman's gentleman". As the years went by, McDonald came to trust Budlong's judgement to such a degree that when Budlong suggested changes in the Zenith line, McDonald ordered them with almost no regard to tooling costs or short-term lost production.[4]

Research efforts have not yet established exactly which pre-war Zenith radios were designed by industrial designers and which were not. Fragmentary evidence indicates that Budlong designed a number of the Zenith consoles as well as many, if not all, of the pre-war bakelite table models. Judging from his post-war work for Zenith, he almost certainly designed the now classic 1941 "Poket Radio," along with a number of the other Zenith pre-war portables. It is also almost certain that he designed some, if not all, of the models of the Zenith "Universal" portables, the immediate precursors to the Trans-Oceanic. After the war, and until his death in 1955, Budlong and his firm were the designers of virtually the entire Zenith line.

Industrial Designer Robert Davol Budlong in his office at 333 N. Michigan Ave., Chicago. c. approx. 1940. *Courtesy of the Budlong family.*

The Zenith Universal Portable

The term "universal portable" was a generic term used both before and after World War II for a portable radio which would play on 110 volts AC or DC house current, as well as on dry batteries fitted within the case. Another term used at that time for a multi-powered radio was "three-way." This type of portable was introduced by most major American radio manufacturers in their 1940 line and was primarily based on the recently developed low drain 1.5-volt loktal tubes. When Zenith introduced its 1940 line of three portable models, the most expensive was named "The Universal"; this name was carried by the most expensive Zenith AM portable each year for many years to come. The very first Universal, the 1940 Model 5G401, contained the newly-patented Wavemagnet and was the model pulled from the assembly line to serve as the test-bed prototype for the development of the first Trans-Oceanic. The 1941 line of Zenith portables was much ballyhooed, with the top-of-the-line Universal, Model 6G501, being called **"The World's Greatest Portable Radio-Guaranteed to Play Where Others Fail or Your Money Back."** Like almost all other portables of the day, Zenith Universals looked like small suitcases when closed for travel. Research suggests that Robert Davol Budlong was consulted on the appearance of both the 1940 and 1941 Universals.

The 1942 Zenith Universal portable, Model 6G601, was available in four finishes, Blue-Gray Airplane Fabric (6G601D), Brown and Ivory Airplane Fabric (6G601MH), Genuine Top Grain Cowhide (6G601L) and Brown Alligator (6G601ML). The Brown Alligator version is certainly the most elegant portable radio produced before World War II; it became the "companion" AM portable to the first Trans-Oceanic. Again, judging from his post-war work for Zenith, the exterior of the 6G601 Universal was most probably designed by Robert Davol Budlong.

1942 Zenith Universal Model 6G601.

Built Like a Fine Piece of Luggage

7 POINTS OF SUPERIORITY

1. **LIGHTWEIGHT—BALANCED WEIGHT**—Can be carried in one hand wherever you go. Case scientifically designed so that weight is equally distributed.
2. **BUILT-IN WAVEMAGNET**—Zenith's newest and most sensational radio improvement . . . assures remarkable sensitivity . . . no aerial or ground needed.
3. **ON AND OFF INDICATOR**—A visual indicator that tells when the radio is on or off—a safety feature.
4. **COMPLETE CASING**—Cabinet closes completely when not in use. Dial and controls protected against damage in transit.
5. **BEAUTIFUL TONE**—Big full tone 5½ inch speaker . . . same beautiful tone for which all Zeniths are famous.
6. **MODERN SUPER-STYLING**—A complete choice of finishes to match popular types of luggage.
7. **GUARDIAN REMINDER**—The battery pack safeguard that says "Turn me off when you are through." Never a chance of wasting "juice" by accident.

Guardian Reminder

SPECIFICATIONS

Powerful, sensitive and selective 4 tube superheterodyne with six tuned circuits. The tubes are the long life, low drain type. Cases are smartly styled with full sized, strong convenient handles, and modern hardware. Size 9³⁄₁₆" high x 11⅝" wide x 8" deep. Single wave band 540 K.C. to 1650 K.C.

Built-In WAVEMAGNET

Brings in the programs alive without aerial or ground. Zenith Portables will operate equally as well in your home, boat, automobile, or aboard train . . . All this freedom of use and adaptability is provided because of the Wavemagnet—the newest and most advanced radio feature. Another Zenith FIRST.

Single Unit BALANCED POWER PACK

*250 HOURS OF SERVICE

A powerful, long life battery, especially designed for portable radio use. Hand selected cells that are balanced scientifically to the radio—assure you many happy hours of entertainment at a minimum cost.

* 250 hours guaranteed on basis of 25 hours a week operation. The power pack fits snugly and securely inside the cabinet. Power pack replaceable anywhere. Individual batteries can be substituted without cutting the wires.
No. Z-59 . . . . Price **$3.75**

Form No. 92OR—4   Printed in U.S.A.

THE NEW 1940 ZENITH PORTABLE RADIO

WITH THE SENSATIONAL BUILT-IN WAVEMAGNET
. . . NO AERIAL . . . NO GROUND

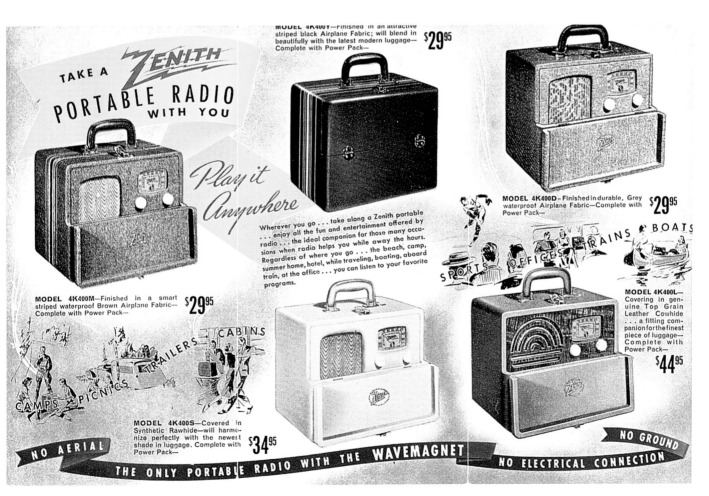

The 7G605 Trans-Oceanic Clipper

In late 1941, plans were well underway to introduce the 7G605 Trans-Oceanic Clipper which had, as we now know, an upgraded Universal chassis. Zenith made the decision to add the new Trans-Oceanic to the line some months after the rest of the line (including the Universal 6G601) was set and early catalogs had been printed. Only the very last 1942 dealer brochures and pamphlets show the Trans-Oceanic. It is pure supposition, but still very likely, that the two year development effort on the world's first shortwave portable was not yet complete when the remainder of the 1942 line was introduced. Comparing the appearance of the 1942 Universal and the Trans-Oceanic Clipper, it is very evident that the appearance of the Clipper is a direct, and somewhat clumsy, modification of the finely proportioned Universal. The cabinet was extended laterally several inches and relatively crude metal escutcheons were added for bandswitch buttons and tone controls. Both portables originally featured a sloop-rigged sailing yacht embroidered in white thread on the black open-weave speaker grille cloth. Very soon after the Japanese attack on Pearl Harbor, during the second production run of the Clipper, Commander McDonald personally ordered that the sloop on the 7G605 Trans-Oceanic be replaced with a generic World War II-style, four-engine bomber; both versions, however, are referred to as the Clipper.

From an engineering point of view, of course, the 7G605 Clipper was completely innovative. It was, after all, the world's first all-wave portable radio. The achievements of the Zenith engineering design team, however, overshadow one singular contribution by Commander McDonald, himself: the multi-band bandspread dial. This dial, which focuses on tuning the International Broadcast Bands, set the Trans-Oceanic apart from almost all other shortwave portables ever produced and made a major contribution to the longevity of the line. A 1951 memo from the Commander illuminates this significant decision:

> I sent a radiogram from my yacht to our engineers telling them to produce a combination broadcast and shortwave receiver, with split bands to make tuning easy. The little portable that I had with me required micrometric fingers to tune.[5]

The Commander was referring to an incident in the summer of 1939, and the "little portable" that he mentions was one of the early shortwave-only portable prototypes that was based on the 1940 Zenith Universal, similar to the one described by Howard Lorenzen as the first (of some 20) prototype.[6]

The Post-War Trans-Oceanics

Commander McDonald tried several times in the early years of their association to persuade Robert Davol Budlong to become a full-time Zenith employee and head an in-house Zenith industrial design group. Budlong demurred, wishing to continue to design Zenith radios, while maintaining his freedom as a consultant to a number of other non-radio manufacturing concerns. In 1938, however, Budlong did move his office to the well-known 333 North Michigan Avenue Building.[7] The ground floor of this structure housed the corporate showrooms of Zenith, showcasing, but not selling, the entire Zenith Line. This showroom proved invaluable as a reference center for Budlong and his staff over the years.

As the war drew to a close, Zenith, along with other American manufacturers, anticipated shifting from all-out war production to civilian production. They looked forward to years of excellent sales to satisfy pent-up consumer demand. Some evidence indicates that Budlong was responsible for the industrial design and styling of most of the post-war Zenith line. In fact, he and Zenith began designing what was to be the 1946 line in the spring of 1944. Soon after the war, Budlong's workload was such that he needed to add other designers to the firm; his staff was never large--three or four people total. In addition to Budlong, the staff consisted of Ken Shory, a graduate of Pratt Institute in New York; Julian Greene, a graduate of the Art Institute of Chicago; and Dana Mox, a graduate of the Institute of Design (Chicago).

Budlong and his associates favored the colors of black or dark brown with bright gold trim for the Zenith line. This color scheme was used throughout the 1940s and most of the 1950s, and related at least in part to the Zenith motto of the era, "The Royalty of Radio." A review of the immediate post-war Trans-Oceanic (8G005Y) and its Universal (6G001Y) companion reveals the design theme for most of the post-war Zenith line: the use of concentric semi-circles and other related forms. This late Art Deco design theme, along with the use of full-circle design motifs, is found throughout the Zenith line for the remainder of the tube era.

8G005Y Trans-Oceanic

G500 Trans-Oceanic

H500 Trans-Oceanic

600 Series Trans-Oceanic

One of the strongest memories of people involved in the design of the Trans-Oceanic series was the personal involvement of Eugene McDonald in each model's development. The Commander worked with the Zenith engineering staff during the development of each new chassis and was intimately involved with reviewing the industrial designers' work from early "concept sketches" to the completed final mock-up prototypes.

Suggestions for the major changes and upgrades of the tube-type Trans-Oceanics came from a variety of sources. Some, especially technological advancements, came from the Zenith engineering staff. Other ideas emanated from the sales force. Many probably also came from the Commander and his yachting friends, who were often featured in the Trans-Oceanic advertising of the day. Most of the ideas for cabinet upgrades, materials changes and styling, of course, came directly from Robert Davol Budlong and his staff.

When a new Trans-Oceanic model was being prepared, Budlong's staff would first prepare "concept sketches," usually in the form of lightly-rendered freehand drawings on tracing paper. These were reviewed by Zenith staff members and by the Commander himself. The Zenith responses to these sketches would then guide the designers as they developed the final design, and then, the layout drawings. The final design was usually presented to Commander McDonald in the form of a full-scale model. Budlong, along with associates Shory, Greene and Mox, designed all of the tube-type Trans-Oceanics and most of the remainder of the Zenith radio and TV line from 1945 until Budlong's death in 1955.

Both Dana Mox and the Budlong family relate that the relationship between Budlong and Commander McDonald became so close that, in September 1954, McDonald sent Budlong to his own personal physician at Johns Hopkins in Baltimore seeking medical advice to stem Budlong's mysteriously-deteriorating heart. Within 24 hours, Budlong was diagnosed with congestive heart failure and treatment was begun. Unfortunately, the damage to his heart was done and he died at, age 53, on February 13, 1955. After Budlong's death, the firm was reorganized as Ken Schory Associates and continued in industrial design practice.[7]

A tremendous volume of excellent industrial design was produced during the 20-year life of the Robert Davol Budlong firm. Numerous designs were executed for the Victor Cash Register and Adding Machine Company, Sears, Sunbeam, Amana and other major midwestern manufacturers. Beside the many classic Zenith radio designs, the Budlong office produced two other designs now considered classics of 1950s design: the original Sunbeam ShaveMaster shaver and the absolutely classic all-chrome Sunbeam two slice toaster.

Sunbeam toaster as designed by R.D. Budlong.

Boldt and Associates

With Budlong's death in 1955, the Commander and senior Zenith executives sought another industrial design consulting firm to aid in designing Zenith radios and televisions. In rather short order, they selected Mel Boldt and Associates. This relatively new firm was then doing excellent design work for a number of major manufacturers, including radios and other appliances for Crosley.

When Zenith retained Boldt and Associates, the firm was less than four years old as a design consulting firm. Mel Boldt's first position as a designer had been as part of an advanced "think tank" design group at The Bendix Corporation. This group was looking at the far future of home appliance design and developed, among other things, the modern Bendix washing machine. About 1950, Bendix reorganized their product design process and Mel Boldt decided to open his own consulting practice on Michigan Avenue in Chicago. The new firm's first account was a continuation of Boldt's in-house work with Bendix. By the time they were retained by Zenith in 1956, Mel Boldt's office had eighteen working designers. In addition to the Bendix account and their work with Crosley, they were working on designs for Fedders Air Conditioning, Auto Point Pencils, American Cabinet Hardware (a juke box) and National Presto. In 1956, they were designing the now well-known totally-submersible Presto Electric Skillet with the removable control.[9]

During the next thirty years of practice, the Zenith account was certainly one of the most important to the Boldt organization, but they continued to produce designs for many other well-known manufacturers. These included the Schick Corporation (electric shavers), Bausch and Lomb (ophthalmic and fashion sunglasses), Amana (refrigerators and other home appliances), the Cushman Company (all-terrain vehicles), Moen Products (faucets) and many others. By all accounts, working at Boldt and Associates was a heady but very demanding experience. The designers were expected to meet almost impossibly short schedules and very high quality standards.

The direct contact between Mel Boldt and Associates and Zenith was handled either by Boldt himself, or by the Director of Design in the office, Mr. David Chuboff. Chuboff spent considerable time at the Zenith Engineering Department working with the staff before projects were brought to the Boldt office for design of cabinetry and hardware.

Anthony J. Cascarano, the actual designer of both the Royal 1000 and Royal 7000 transistorized Trans-Oceanics, was a very highly regarded young designer in the Boldt office when the Zenith account came to the firm. After graduating from the Art Institute of Chicago, Cascarano had worked in the famous Ford Styling Department in Detroit, and for International Harvester, before joining the Boldt office in 1955.[10]

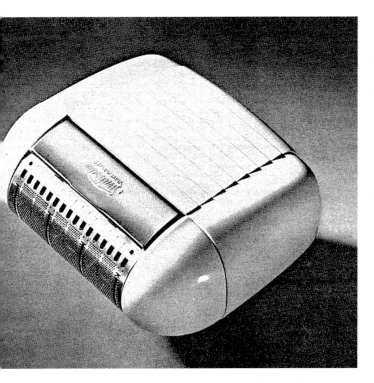

Sunbeam Shavemaster as designed by R.D. Budlong. *Courtesy of the Budlong Family.*

Today, living in retirement in Lake Forest, Illinois, Tony Cascarano fondly remembers his sixteen years with the Boldt office:

> Boldt's office was quite a fabulous place to work, because the design staff was a very enthusiastic bunch of people. Thanks to Mel's contacts and the reputation of the office, the kind of projects that we worked on would have been prized in any design office. A high percentage of the projects were consumer goods, housewares, etc.

When discussing the aesthetics--the "look"--of the era, Cascarano related:

> Industrial design of that period [1957] was beginning to turn to the "clean look." That trend required more concern be given to the relationship of materials, to proportions and to the relationship of the design elements to each other. Controls were not treated just as decorative parts, but were designed to create an overall functional look and at the same time provide an aesthetically desirable "high-tech" look. I suppose that the influence must have come from architecture, Mies van der Rohe and the others of the International Style. Raymond Loewy [the most prominent industrial designer of the 1950s and 1960s], whose Chicago office had both Hallicrafters and Studebaker as clients, was probably the main influence in industrial design at that time, especially in Chicago. If we wanted to see what was new in the industry, we usually went to the Merchandise Mart. There was also a showroom on the west side of Chicago that displayed communications receivers, some of which were designed for Hallicrafters by Loewy.

Tony Cascarano also related his memories of Zenith's corporate commitment to excellence, especially in audio quality, and to state of the art design in all aspects of their products. [By this time, the corporate slogan had been changed to read, "The Royalty of Radio *and Television*."] Cascarano and others of the design community still recall the elegant atmosphere of the large Zenith corporate showroom at 333 North Michigan Avenue in Chicago. It was a very large showroom, and each product was beautifully displayed. The furnishings and several pieces of sculpture in the Zenith showroom were designed by the world-renowned sculptor, Harry Bertoia. This corporate commitment to excellence and to contemporary design had a major impact on the designers creating the products of Zenith's future.

During his years at Mel Boldt and Associates, Cascarano designed a number of other products. His most memorable designs for Zenith include Zenith radio Models B-513V, J506C and the beautiful AM/FM table receiver, the S-46353. For Presto, Cascarano designed the "Rock and Mix" vertical handled portable mixer and the "Hotdogger" hot dog cooker. Cascarano is particularly remembered for his design of the Moen single handle faucet for home and institutional use. He also designed major home appliances for Amana and Norge.

The Design of the Royal 1000

As the two-year-long engineering effort to design the first transistorized Trans-Oceanic was nearing completion, the industrial designers were called in to discuss the design of the new cabinet. Since the transistorized Trans-Oceanic represented a technological revolution, a completely new aesthetic approach was appropriate. The Boldt office representatives at this important meeting were, most probably, David Chuboff and Mel Boldt.

In this first design conference with the Boldt organization, Commander McDonald related his recent purchase of a German-made Leica camera. He rather strongly suggested that the black leather and brushed chrome of this beautiful new high-tech camera be considered as cabinet materials for the new Royal 1000.[11]

Tony Cascarano relates that in the case of the Royal 1000, the Boldt staff was conscious that the Trans-Oceanic was "Commander McDonald's baby"; however, it was handled relatively routinely in the design studio. By the time the project reached Cascarano, the chassis was basically designed; the new innovative rotary dial/bandswitch combination, an idea of the Zenith engineers, was already in place. The industrial designers were, however, permitted to shift the position of the volume, tone and tuning controls slightly.

After the initial design meetings, Cascarano remembers developing a number of design concept sketches for several possible cabinet designs. These were discussed with Zenith and were almost certainly reviewed very closely by Commander McDonald himself. From that process, a design direction for the Royal 1000 cabinet was determined. Cascarano then made a very precise full color rendering of the Royal 1000 in pastel chalk. This drawing was shown to Zenith and used to develop a final full-scale mock-up prototype that was developed under the direction of Chester Wojtowicz of the Boldt office.

The Royal 1000 design was not a rush job and was "on the board" in the design office for several months; things did get a bit rushed at the end, however. Both Cascarano and Wojtowicz remember Boldt standing around on "The Day," waiting anxiously for the staff to put final touches on the full-scale mock-up.

As the last drop of paint dried, Boldt took the mock-up and headed for the corporate offices of Zenith on Chicago's west side. Arriving there, he proudly presented the model to the Zenith executives, letting them know that it could finally be taken up to Commander McDonald.

Probably because of their respect for McDonald's high standards and his gruff demeanor, the Zenith executives gave the model back to Boldt and said, "No Mel, *you* take it up to the Commander!" Having little choice, he did just that, only to return later, all smiles. The Commander had looked over the design mock-up very carefully and then said "That's *exactly* what I wanted!" The first presentation prototype was precisely what we all know today as the Royal 1000 Trans-Oceanic.[12]

Apparently, Cascarano and the Boldt design staff were not conscious that they were in the process of creating both an industrial design and a radio classic. He and the firm were much too modest to think in those terms. They, however, were certainly acutely conscious that they were designing a very special radio for a very special client.

Royal 1000 Trans-Oceanic

The Royal 2000

The Royal 2000 was not, as many believe, a member of the Trans-Oceanic series. It was, however, a milestone in radio history: the first American AM-FM portable radio.[13] Like most of the rest of the Zenith radio and TV line of the solid-state era, it was designed by the team at Mel Boldt in conjunction with the engineers at Zenith.

The introduction of an FM portable lagged far behind the introduction of FM itself. In the first two post-war decades, virtually all FM stations were relatively low-powered and almost all featured classical rather than popular music. For these reasons, radio manufacturers decided that there would be little or no interest in portable FM radios. By 1960, Zenith concluded that the time was right to introduce a large "lunch box" portable radio with FM as well as AM coverage. Since both the Zenith radio line and FM itself were known for excellent audio fidelity, the Royal 2000 was designed essentially as a large heavy speaker enclosure containing an excellent 5" x 7" oval speaker. To clear the center of the cabinet for use as a sound box, and to clear the very large speaker magnet, the chassis of the Royal 2000 was turned upside down and hung from the underside of the cabinet top. For acoustical reasons, there was no attempt to miniaturize either the cabinet or the chassis, and no attempt was made to make the Royal 2000 a featherweight. The final configuration of the Royal 2000 was 12" x 8" x 5" deep and weighed in at twelve pounds.

Gordon Guth, then a young industrial design graduate from the University of Illinois, Champaign-Urbana, was given the task of designing the Royal 2000 as one of his first projects for the Boldt organization. He was asked to develop a design which would be able to join the then emerging family of solid-state Trans-Oceanics.[14] Guth did a wonderful job. In addition to the classic early 1960s cabinet by Guth, modern owners of the Royal 2000 continue to enjoy some of the mellowest audio ever to come from a solid-state portable radio.

Royal 1000 Trans-Oceanic (closed) and Royal 2000 Trans-Symphony.

The Royal 3000

From both an electronic and product design viewpoint, the Royal 3000 was simply a "face-lift" of the Royal 1000. Electronically, the FM sub-chassis was added. The Boldt organization redesigned the front panel, maintaining the same front door and dial escutcheon, and sculpted the new bulged rear door necessary to contain the deeper chassis. This was not a memorable project and probably took no more than a day or two of one of the senior designers' efforts.

Royal 3000 Trans-Oceanic

The Royal 7000

By the late 1960s, the Royal 3000 had become obsolescent and a number of other major manufacturers had entered the field. Contemporary evaluations of the all-wave portable field were published in 1967 by *Popular Science* magazine and *Consumer Reports*.[15] The field of all-band portables had become quite crowded; there were 11 major portables competing for the rather limited market. These were clustered in two price groupings: four receivers costing from $200 to $300, and six receivers costing about $100 each. The then decade-old Royal 3000 was priced at $199.95 and was rated clearly inferior to the $299 Panasonic RF50000A and the $239 Grundig Satellite TR-5000.

Royal 7000 Trans-Oceanic

By 1967, the loss of market and the technical obsolescence of the Royal 3000 must have been no surprise to Zenith executives. After all, the Royal 3000 was merely a face-lift of the original Royal 1000, which was designed at the dawn of the solid-state era.

At this late date, there is neither adequate information nor need to place blame for the diminished stature of the Trans-Oceanic. Many factors undoubtedly played a role in the affair. Certainly, one contributing factor was the fact that Commander Eugene McDonald had died in 1958 and was no longer there to lavish attention on "his" radio. The ever-decreasing role of consumer radios in the manufacturing mix of Zenith surely played a significant role as well.

By late 1967, Zenith executives decided to design a new Trans-Oceanic, the Royal 7000. The electronic design and basic chassis were, of course, designed by the Engineering Department. Mel Boldt and Associates was responsible for the final cabinet configuration and all of the visible detailing of the receiver. A then mid-career Tony Cascarano was responsible for the design work on the Royal 7000, as he had been on Royal 1000 a decade earlier.

There were many improvements in the electronic design of the Royal 7000, including switchable (wide-narrow) selectivity and a BFO (beat frequency oscillator) to allow reception of code telegraphy and SSB (single side-band) voice communications. Zenith engineers chose *not* to adopt the then state of the art printed circuit board technology in use in many other parts of the electronics industry. This decision to remain with a heavy metal chassis, point-to-point wiring and individual transistors was based on the belief that the rugged nature of metal chassis design, and the substantial size and weight of such construction, would continue to be important to potential Trans-Oceanic purchasers. Unfortunately, Zenith and some other American manufacturers read the trends incorrectly: American consumers flocked to the smaller, much lighter electronic products from east Asia. Equally unfortunately, the point-to-point wiring locked production of the Royal 7000 into a labor-intensive, hand-assembly process.

For most customers the many electronic innovations of the Royal 7000 were overshadowed by the many improvements to the exterior of the radio; the proportions, the materials and the colors of the Royal 7000 cabinet were absolutely elegant. Then, as now, the styling of the Royal 7000 set it apart from its competitors, both foreign and domestic. Aesthetically, the Royal 7000 is in a class by itself.

There were numerous practical improvements in the design of the Royal 7000 as well. Most importantly, the Waverod antenna was made entirely separate from the carrying handle. This enabled the radio to be easily lifted while playing, and eliminated the part most likely to break on the Royal 1000/3000 cabinets (the combined handle and Waverod). The door protecting the front panel of the radio was also totally redesigned. Whereas the Royal 1000/3000 door was a single-hinged, two-pieced door, the Royal 7000 had two doors. The lower door dropped down as before, but now could slide horizontally back into the cabinet bottom. This position was both out-of-the-way and stable for moving the radio while in use. The upper portion of the front door was now L-shaped in cross section and hinged at the center of the cabinet top. It was designed so that it could rotate open, like the old front doors on the tube-type Trans-Oceanics. A beautiful world map and time chart were positioned where the Wavemagnets had been on previous models. Below the world map, but integral with the front door, was a continuous loop of plastic with silk-screened notations of the hours of the day/night. This loop, controlled by two thumb wheels, could be properly positioned below the map so that the correct local time throughout the world could be read instantaneously. This world map and time line are considered by many to be the most useful of any ever produced on a portable radio.

Who exactly suggested each of these improvements is not now known. Certainly a number of the practical problems with the Royal 1000/3000 were pointed out by Zenith-equipped expeditions[16] and by stateside users. Some of the design ideas embodied in the new Royal 7000 undoubtedly came from the engineers at Zenith; others must have come from the creativity of Cascarano and his colleagues at Mel Boldt and Associates. Wherever these ideas came from, Zenith and Mel Boldt and Associates wove them into a classic of late-1960s Americana. Today, the Royal 7000, Royal 7000-1 and the Royal D7000Y are as pleasing to the eye and ear as they were when new and continue to serve their proud owners faithfully.

The Last of the Line — The R-7000 Trans-Oceanic

Sometime in late 1977 or early 1978, Zenith executives and engineers decided to design a new member of the Trans-Oceanic line. Sony had begun "successfully" marketing a top-of-the-line portable *with a digital frequency read-out*: the Model CRF320A was introduced in 1976 at $1495.[17] Rather than continue the long tradition of being the best and most expensive all-band portable on the market, Zenith chose to design the next Trans-Oceanic as a "specialty item" radio with a production target of 20,000 units per year.[18] Instead of positioning this specialty item at the top of the market opposite the Sony, Zenith executives targeted the retail price near the middle price bracket, at $300 to $350.

The electronic design of the new Trans-Oceanic came from the Radio Group within the Engineering Department at Zenith. The lead designer of the R-7000 was Larry Latta; Don Shoop did the design of the VHF and FM tuners. A young engineer, Robert Stender, assisted both and continued to solve production-related problems during the three-year life of the R-7000.[19]

The electronic design of previous Trans-Oceanics had always focused on excellent reception of foreign broadcast programming, accomplished by devoting five or six full bands (electrically bandspread) to the relatively narrow International Broadcast bands and two "general shortwave coverage" bands. This focusing on the crowded International Shortwave Broadcast bands was an attempt to solve "the dial problem." The "problem" had always been the difficulty of finding a particular station, even if its frequency was known, and returning to a station once it was tuned out. Spreading one narrow broadcast band over 6 or 8 inches of the dial, as had been done on all previous Trans-Oceanics, was the only known solution to this problem prior to digital frequency read-outs.

R-7000 Trans-Oceanic

Probably due to the cost of the LED digital read-outs in the late 1970s, Zenith decided *not* to incorporate this relatively new technology in the newest Trans-Oceanic. They also chose to spread the shortwave spectrum evenly throughout the six dials available, rather than concentrate on the International Broadcast Bands; this was probably in hopes of attracting a wider spectrum of radio enthusiasts to the *new* Trans-Oceanic. They also included one electrically-spread shortwave band: the 27 MHz Citizens Band. The decision to favor general shortwave coverage relegated the previously 6-to-8-inch-wide International Broadcasting bands to dial segments of less than one inch each! This change in Trans-Oceanic design philosophy was almost certainly a sales, rather than an engineering, decision.

Tuning stations on the two general coverage bands that were on most Trans-Oceanics had always been very difficult. Now that all bands were general coverage, the Zenith engineers developed a concentric dual tuning knob. The outer knob tuned at the normal rate, while the inner knob was for fine tuning; it tuned at a rate 40 times slower than the outer knob. This "magic" was originally accomplished by what turned out to be a complex belt drive system. The belt drive tuning system was used during the first two years of R-7000 production; it was not highly regarded, primarily due to the "backlash" of the somewhat flexible belts. This backlash made tuning the now general coverage shortwave spectrum even more difficult. The tuning system was changed to a much more successful gear drive system in the final year of production. Although the gear drive system eliminated backlash, the difficulty of finding a single shortwave station on a known frequency remained because of the general coverage analog dial.

In keeping with the long tradition of rugged excellence, the engineering team chose to base the R-7000 design on military-grade epoxy-glass printed circuit boards and to select key components (such as the bandswitch) from the same class of components. In fact, much of the physical design of the interior of the radio utilized concepts and technology, such as plug-in boards, developed in the military and space electronics field.

There were numerous electronic improvements in the new design. The most important of these included two large tuning and signal strength meters, a switchable Automatic Noise Limiter for AM and SW and a switchable Automatic Frequency Control for FM. The switchable bandwidth selectivity circuit of the Royal 7000 was retained and vastly improved.

In an attempt to curb costs and relate to the Trans-Oceanic tradition, Zenith executives decided to retain the cabinet design of the now decade-old Royal 7000.[20] They asked Mel Boldt's office to do a thorough face-lift of the old cabinet and to redesign the front panel to incorporate the new controls and meters. The project was assigned to a highly-regarded industrial designer at Boldt's, Rick Althans. Althans did a wonderful job of composing the numerous controls in a both practical and beautiful arrangement. That, in itself, was no mean feat since the four front controls (tuning, tone, volume and dial light) of the Royal 1000 had grown to eleven on the new receiver! Althans also changed the color scheme of the receiver from the Royal 7000s dominant chrome to a predominantly black look. Many collectors consider the all-leather and gold slide-rule dial model 600 of the tube years and the Althans designed R-7000 to be the two most beautiful of the Trans-Oceanics--especially poetic since each represents the end of an era.

The R-8000 Trans-Oceanic

The Zenith Electronics Corporation chose to withdraw from the radio market in 1982. At the time of this decision, the Radio Group was at work on the design of yet another generation of Trans-Oceanics! This new design, incorporating digital frequency read-out, never progressed beyond the earliest engineering prototypes. In fact, although the engineering team referred to the design informally as "the R-8000," Zenith had not yet assigned an official model name to the new set when all radio design work was terminated.[21]

The Tradition of Excellence

From start to finish, the Trans-Oceanic line was marked by excellent design. Electronically, the early models were innovative designs which broke new ground in consumer electronics. The later years of the line still exhibited excellent electronic design, though these models were less innovative than their forebears. The aesthetics--the industrial design of the Trans-Oceanic--were excellent, from start to finish. The beauty of these radios is one attribute which separates them from almost all of their rivals. This grace of styling was partly due to the fact that Chicago was one of the world centers of post-war industrial design. The lasting beauty and the electronic qualities of the Trans-Oceanic line are also directly attributable to Commander Eugene McDonald, Jr. The Commander selected excellent designers and then demanded that they give their very best.

'ROUND THE WORLD TOUR.

Give your ears a vacation, with the radio that's powered to tune in the world. Eleven-band reception, including FM, AM, long and short wave, marine, and weather bands. Runs on "D"-cell flashlight batteries or plugs into any 115- or 230-volt AC outlet. Includes built-in antennas, earphone and jack, flip-up time-zone map, and log chart listing world station frequencies from Poughkeepsie to Peking. Hear The Trans-Oceanic® portable, model D7000Y, at your Zenith dealer.

ZENITH®

The quality goes in before the name goes on.®

Chapter 5

THE END OF
THE TRANS-OCEANIC
An American Tragedy

In the early 1980s, Zenith entered a ten-year relationship with the Heath Company. Part of that arrangement included marketing the model R-7000 Trans-Oceanic in the HeathKit catalogue. The dismal sales performance of the R-7000 when sold by Zenith apparently continued under the Heath banner. Soon, internal memoranda were circulated to all employees of the Heath Company offering this almost $400 receiver for about $150; the same memos were circulated to all employees of the Schlumberger Corporation, the previous owner of Heath.[1] The final retail appearance of the once-vaunted Zenith Trans-Oceanic was on the shelves of several midwestern discount houses where the last of the R-7000s were dumped for about $50 each![2] For those who love radio in general and the Zenith Trans-Oceanic in particular, it is almost impossible to believe that the Trans-Oceanic line could have ever come to an end. It did end, and it was a distinctly American tragedy.

Viewed closely, the history of the Trans-Oceanic is a nearly perfect model for the history of the entire American consumer electronics industry. Some of the seeds of this demise can be found in global trends in electronics manufacturing, some lay in the history of Zenith, while others fall to the cycle of the careers of particular individuals involved in the design and manufacture of the Trans-Oceanic. At each level of focus, particular trends, and even particular individuals, had to be in the right place at the right time for the Trans-Oceanic to be born and prosper. Conversely, changes in these trends and the loss of several important individuals spelled an almost certain end to the Trans-Oceanic, no matter what decisions were made by particular Zenith executives in the late 1970s and early 1980s.

The Industry

Churchill and many others have observed that there are tides in human affairs. There certainly have been massive tides in the global consumer electronics industry. The thirty years between 1935 and 1965 mark the era of global dominance by the corporate giants of American consumer electronics. Those years also encompass the Trans-Oceanic's finest years.

The electronics corporations of Japan grew out of the ashes left from the American bombing raids of 1944-1945. Although basically starting from scratch at war's end, by the 1960s the Japanese electronics industry was blessed with a comparatively new industrial infrastructure and with new modern plants financed primarily by U.S. capital. They were also doubly blessed with a relatively compliant work force and a 2500 year-old culture which thrived on almost endless refinement and incremental improvements of manufactured objects. Finally, Japanese corporations had developed a corporate and financial structure which rewarded long-term success, rather than the American focus on quarterly or annual profit and loss.

In retrospect, these and other factors made the rapid decline and near-demise of the American consumer electronics industry virtually inevitable. Motorola survived and prospered by shifting almost totally to the manufacture of business and governmental communications equipment. RCA moved most manufacturing "off-shore" and is, today, a shell of the proud giant of the 1950s and 1960s. A smaller Zenith Corporation survived by persistence, dedication to quality, and by abandoning the radio market entirely in the early 1980s. Philco, Admiral and Crosley are no more.

The Circumstances

By examining the volume and content of Zenith magazine advertising of the post-war era, it is rather easy to discern the Zenith commitment to the radio line and even to the Trans-Oceanic itself. Radio was King in the early post-war era, and the Trans-Oceanic was advertised heavily both in general circulation magazines and in travel-oriented journals like *Holiday* and *National Geographic*. The amount of advertising for the radio line and for the Trans-Oceanic seems to have remained rather constant until 1960. The Royal 1000 was introduced in late 1957 with a publicity blitz that rivaled any other major Trans-Oceanic model introduction and was advertised heavily through 1960. After 1960, however, print advertising for the Trans-Oceanic virtually ceased except for a few scattered advertisements in radio trade and electronics hobbyists magazines.[3]

Taken in total, the pattern of Zenith general circulation print advertising in the period 1960-1983 illustrates the ever-increasing role of television as the major product of the Zenith Electronics Corporation. The pattern also probably indicates the increasing role of television as the major American advertising medium, since the total number of radio product print advertisements dropped steadily throughout the 1960s and 1970s. However, within the radio-oriented print advertising that was purchased, the Trans-Oceanic line was noticeable by its absence. There were years when there was more advertising for the Motorola and Admiral "look-alike" portables than for the Trans-Oceanic. This extraordinary circumstance would never have happened "while the Commander was in charge." It is interesting to note, however, that several TV commercials for the Zenith Trans-Oceanic were produced by Foote, Cone and Belding, the large ad agency long retained by the corporation. These were produced in 1967 and 1968 for network use. Nevertheless, the general trend in Zenith advertising in the 1960s and 1970s was away from advertising radios in general and Trans-Oceanics in particular.

Following the Commander's death in 1958, the Trans-Oceanic seems to have languished. Previously, when new electronic technology reached the consumer field, a new Trans-Oceanic was not far behind. When introduced, this new Trans-Oceanic was clearly superior to the rest of the field and was usually the most expensive portable radio in the world. This pattern had been true from the beginning of the line. It was also true at the introduction of the Royal 1000 in the 1958 line (over $1,300 in 1994 dollars) and the Royal 3000 (also over $1,300 in 1994 dollars).

The Royal 3000 was clearly held on the market too long (November 1962 until the June 1969 introduction of the Royal 7000). By 1967, reviews of the all-wave portable field in *Consumer Reports* and *Popular Science* portray the Royal 3000 as dowdy and old-fashioned. Its $200 price was in the lower part of their upper groupings. The largest, best and most popular was the very modern Panasonic RF-5000 offered at a top-of-the-market $300. Zenith continued to sell the Royal 3000 for another two years.[4]

When Zenith was designing the Royal 7000 in the late 1960s, elements within the electronics industry held two widely differing views of the future buying preferences of U.S. consumers. One view held that the then new printed circuit technology would be welcomed by consumers with open arms and wallets; it was felt that printed circuit technology offered wide possibilities of miniaturization and could provide many more functions for less money. The other view held that Americans had always believed that bigger and heavier was better, and that the public would not accept expensive products which were featherweights.[5] It is easy to see how Zenith executives would opt for the large and heavy approach, given the Trans-Oceanic's long history of rugged service in

war and peace. With the new Royal 7000, Zenith brought out a radio with a beautifully designed cabinet and noticeably upgraded circuitry. Unfortunately, their decision to maintain the all metal, hand-wired chassis was a misreading of the market; their decision also locked the Royal 7000 into a costly labor-intensive manufacturing process.

Throughout the 1970s, various versions of the Royal 7000 were produced. Probably in response to intense competition from Japan, Zenith did not raise the price of the Royal 7000 significantly throughout that inflation-ridden decade. Consequently, the real price of the Royal 7000 dropped steadily. At introduction in 1969, the Royal 7000 cost $1,135 in 1994 dollars. During the last model year, 1978, the Royal 7000's effective retail price had dropped to under $700 in 1994 dollars. If it was at all profitable to produce the Royal 7000 with old-fashioned wiring and components at the beginning of its life in 1969, it must surely have been a large-scale losing proposition by the end.

The Royal 7000 is a beautiful radio. It is also a fine all-wave portable. It gave good value for its price and it continues to serve its owners well to this day. Historically, however, a top-of-the-line Zenith product-- be it portable radio, table radio, console, stereo or TV—had always been among the most advanced and premium priced product of its type. The Zenith Royal 7000 did not measure up to its own heritage or to its own corporate culture. It could not possibly dominate the market as its predecessors had done. It was a fine middle-priced all-wave portable radio. Period.

The close-out of the Royal 7000 was followed 18 months later by the introduction of the Trans-Oceanic R-7000, in May 1979. This last Trans-Oceanic used near state-of-the-art technology: military-grade glass-printed circuit boards and integrated circuit chips. However, Zenith executives and designers chose to ignore both the top-of-the-line marketing strategy of the past and the new digital frequency readout technology then in use by the Japanese. It is really irrelevant that the beautiful R-7000 (the "Stealth" Trans-Oceanic) also had some performance and mechanical problems. In truth, the radio was doomed from the start by the failure of executives and engineers to understand the implications of digital frequency read-outs to consumers.

The Competition

Although a few European manufacturers remained in the all-wave portable radio market in the early 1980s, the major competition for the Trans-Oceanic came from the Japanese, primarily Sony and Panasonic. Of these two, Sony quickly developed as a new world leader in innovative consumer electronics products.

By the mid-1970s, Sony engineers realized what Commander McDonald and the Zenith engineers knew in 1940: the major shortcoming of shortwave radio from the consumer's point of view was "the dial problem"--the inability to easily tune a shortwave station. This tuning problem is the very reason that the Commander insisted that the Zenith engineers of Trans-Oceanics provide electrically bandspread dials stretching out each International Broadcast Band to cover an entire dial. This strategy was abandoned in the design of the R-7000.

In the mid-1970s, perceptive Sony radio engineers adopted the then new light emitting diode technology (LED's) and a modified tuning circuit design to totally solve "the dial problem." In this tuning and read-out design, each frequency was displayed as highly accurate digits making the tuning of shortwave radios no more difficult than the tuning a TV. In 1976, only the very stiff retail price of $1,495 (almost $4,000 in 1994 dollars) kept the new Sony CRF-320A from sweeping the field of all competitors. The Sony CRF-320A was a radio that the Commander would have loved--breakthrough technology (as far as consumer products were concerned) offering obviously superior performance at a top-of-the-line price. As previously stated, Zenith radio executives and engineers could not have helped but be aware of this receiver when they embarked on the design of the R-7000.

The major weakness of the Sony CRF-320A (in addition to its price) was the fact that the bright red LED frequency numerals drew a great deal of current. This large current drain required large batteries and implied a large and very expensive radio. Three years later, Sony engineers again called on a new technology: liquid crystal displays (LCD's) to solve the battery problem. These now familiar devices could be formed into alpha numeric read-outs, and most importantly, drew almost no current. Sony executives now saw the chance to bring out a digital portable at a more normal "top-of-the-line" price and did so a mere 12 months after the introduction of the Zenith R-7000. The new Sony ICF-2001 receiver was again truly revolutionary and rather quickly swept all competition from the field. The retail price of the star-crossed Zenith R-7000 was $380 at introduction. The Sony ICF-2001 with digital read-out was introduced for $299. Which would you have purchased?

Five years later, Sony replaced the ICF-2001 with an even more revolutionary ICF-2010. The 2010 offered LCD digital read-out and a completely new type "synchronous" detector which solved much of the remaining fading and audio problems associated with receiving shortwave broadcasts. The 2010 also provided 30 programmable memory channels for storing the frequencies of the consumer's favorite stations. All of these features, and more, were offered for $319 at introduction in 1985. Most relevant to this discussion is the fact that, to date, sales of the Sony ICF-2010 are believed to be well above a million units in the U.S. alone; sales worldwide may approach ten million. The U.S. sales of this single Sony model now nearly exceed the total sales of all models of the Trans-Oceanic. The market was there; circumstances conspired to ensure that successor models of the Zenith Trans-Oceanic would not fill it.

Zenith's decision to cease all radio production ended the Trans-Oceanic line. Although Zenith held on to production of radios longer than other American electronics corporations, by the early 1980s, it was obvious that the future of the corporation lay beyond the bounds of the consumer radio market. It is interesting to note, however, that the Zenith Electronics Corporation continues to pay several thousands dollars per year to maintain worldwide trademark on the name Trans-Oceanic.[6]

Zenith Electronics Corporation

The more the authors have learned about the Trans-Oceanic and about Zenith, Hallicrafters, National and a few of the other giants of American electronics, the more they have come to appreciate how unique the Trans-Oceanic really was. They have also come to appreciate the importance of "corporate culture." The behavior of a corporate giant really is modified by what the people working for that corporation believe about it. In the case of Zenith, the slogans "The Royalty of Radio" and "The Quality Goes in Before the Name Goes On" really did, and do, encourage the people in that corporation to strive for the very best.

Zenith recently celebrated the 75th Anniversary of the first Chicago Radio Laboratory receivers. Today, as the only U.S.-owned manufacturer of color TVs and color picture tubes, Zenith continues to maintain its very high quality standards and competes very effectively due to its industry-leading innovations, strong brand name, and cost position. Zenith is currently a leader in the development of the next generation of digital television technologies: high-definition television (HDTV), interactive TV, and digital cable TV. The company will certainly continue to play a significant role in the design and manufacture of consumer electronics in the years ahead.

Although the company has been licensing its name for certain audio products, Zenith does not appear to have any plans to reenter the radio market. While it is impossible to predict which Zenith product, if any, will reach the pre-eminence of the Trans-Oceanic of the 1950s, the future of the corporation seems very bright.

The Individuals

The story of the birth, life and death of the Zenith Trans-Oceanic is at its essence the story of a number of unique individuals. The Trans-Oceanic *really was* the brainchild of Commander Eugene McDonald, Jr.— a captain of industry and a unique figure in radio history. His personal love of exploration and sailing, coupled with his thirst to hear news from abroad, led directly to the creation of the Trans-Oceanic. His continuing romance with adventure and foreign lands assured that the Trans-Oceanic would remain *the* radio of armchair and real adventurers alike. The Commander's personal commitment to quality and to a marketing strategy that targeted the upper portion of the market (hence, the "Royalty of Radio") almost guaranteed that the Trans-Oceanic would be the top-of-the-line consumer portable radio made in America. The Trans-Oceanic remained very much the personal radio of the Commander until his death soon after the introduction of the Royal 1000. The loss of his personal support and guidance was not the death knell of the Trans-Oceanic, but his loss certainly contributed to its eventual demise.

The role of singular individuals in the life of the Trans-Oceanic line is not limited to the Commander. Zenith's Chief Engineer in the late 1930s was the highly regarded G. E. "Gus" Gustafson. Had not Gustafson been very familiar with shortwave from his background in amateur radio, and had he not gathered a very creative design staff, Zenith would probably not have solved the numerous technical hurdles that stood in the way of producing the world's first battery-powered all-wave portable radios. Members of the Gustafson engineering team were just the first of a number of very creative Zenith engineers who worked on the Trans-Oceanics and who were "real radio men." The Trans-Oceanic was *not* "just another radio," and it was not created by "just another bunch of engineers."

Had not the gentlemanly Robert Davol Budlong formed such a strong bond with the voluble Commander, and had not Budlong and his small crew been so very talented, the Trans-Oceanic would not have been the classic of Americana that it is today. Had not Mel Boldt and his staff, especially Tony Cascarano, arrived on the scene in such a timely fashion, the solid-state Trans-Oceanics would not have been the classics of design that they are. These renowned designers did not consciously set out to design "classics" when they created the Trans-Oceanics, but they knew that they were designing a very special radio for a very special client.

It took a whole chain of special individuals, each playing at or above the "top of his game" to create these very special radios. As these individuals faded from the scene through death or diversion to other design pursuits, the Trans-Oceanic was doomed...doomed to be "just another radio," and doomed to fade from the consumer electronics scene.

Gratitude

There appears to be little real chance that the Trans-Oceanic all-wave portable radio will ever be resurrected. That is as it should be. The Trans-Oceanic was, at its best, the product of an excellent corporation and an excellent collection of professionals, all focusing their careers on designing and building a great line of radios, the Royalty of Radios! The Trans-Oceanic was simply an outgrowth of their genius and commitment.

We should all be deeply grateful that we, so close to the 21st Century, can still so readily enjoy the finest fruits of their labor. There are literally tens-of-thousands of Zenith Trans-Oceanics in use today. More of them are appearing each week from long, careful storage in closets and attics and becoming treasured examples of the best of our electronic past.

Thank you, Commander, and thank you, Gentlemen!

Chapter 6

LINEAGE OF
THE TRANS-OCEANIC

The forty-year-long history of the Trans-Oceanic is a living record of the American consumer electronics industry. When considered as a group, the individual models of the Trans-Oceanic series are each not only a reflection of the corporation and people who made them, but also a fairly accurate reflection of the health and orientation of the American consumer electronics industry at the time of their manufacture. The first Trans-Oceanic, the 7G605 Trans-Oceanic Clipper, was the first mass-produced all-wave portable in the world. As such, its circuitry was very innovative and its components, especially the "loktal" tubes, were at the cutting edge of "ruggedized" electronics of the early 1940s.

The post-war tube models, especially the G500 and H500, reflect the rapid advances made in American electronics during the war years. The 600 Series, produced from 1954 until 1962, represented the very best battery-powered portable radio that a mature, confident industry could produce based on vacuum tube technology. It is interesting to note that the B600 was not only the last of the tube Trans-Oceanics, it was the last American-made tube portable.[1]

The Royal 1000 Trans-Oceanic was not the first *all-wave* transistor portable. That accolade belongs to an otherwise pedestrian Magnavox portable.[2] The 1958 Royal 1000 was, however, the first "serious" transistorized all-wave portable. Zenith's leadership of the consumer radio market continued with the Royal 2000, introduced in 1960, the first AM-FM portable made in the United States. It was not a Trans-Oceanic, but it was designed to be a companion receiver to the Royal 1000 Trans-Oceanic.[3] After the early 1960s, the Trans-Oceanic line, and, indeed, the American consumer electronics industry, began to have to share the center of the world's electronic stage. A generation later, the stage would belong to others.

The Royal 3000 Trans-Oceanic was, and is, a wonderful radio. However, it was really a face lift of the Royal 1000 cabinet containing essentially an upgraded Royal 1000 chassis with an added FM section. At its introduction in the 1963 line, it was a very competitive, but not innovative, receiver; by 1968, the American market contained several European and Japanese receivers that far outperformed this already ten-year-old design.

The Royal 7000, introduced in mid-1969, met the intense performance competition from abroad and may have exceeded it slightly; however, it did so with obsolescent technology. Like the entire American electronic industry, the 1970s was a period of continual loss of market share for the Trans-Oceanic. By the late 1970s, the Royal 7000 was woefully obsolete. The R-7000, introduced in 1979, again met, but did not exceed, the competition, even at the targeted middle of the market. The R-7000 was a new radio and new technology for the Trans-Oceanic (printed circuits and ICs) put into an upgraded version of the old Royal 7000 case. The radio was designed in the United States but was, after the first year, manufactured in Asia. It is important to note that the Zenith Trans-Oceanic R-7000 was the last portable radio to be manufactured in the U.S.[4] Unfortunately, it did not perform well, either technically or competitively. A mirror of the American consumer electronics industry, the R-7000 and the Trans-Oceanic line passed into history in the early 1980s.

The Models (*see also the* Endpiece)

Two of the first questions asked about the Trans-Oceanic line are: how many models were there and when was each produced? It is surprisingly difficult to determine the number of models. If the term "models" means substantially different radios, then, there are six tube models: 7G605 Clipper, 8G005Y, G500, H500, R-520/URR (Military) and the 600 Series. If one wishes to include the individual "models" of the 600 Series, the L600, R600, T600, Y600, A600 and the B600, then, there are 11 tube models. All of the "600 Series," however, are substantially the same radio. The differing alphabetical prefix to the model number is simply a Zenith model nomenclature that was adopted in 1955; in this new system, the letter prefix changed once or twice each year, whether the radio changed or not.

If one includes "sub-models," the picture is even more complex. The post-war 8G005Y had two sub-models or versions beyond the initial model: 8G00YTZ1 and 8G005YTZ2. The TZ1 and TZ2 nomenclature connotes successive changes to the power supply in the latter years of that particular model. Similarly, there is a "sub-model" which ran throughout the 600 Series models. There was a leather-covered version of each of the 600 Series models. These brown leather-covered Trans-Oceanics were connoted by the suffix "L" after the model number. The chassis number was also different (6n40 vs 6n41) to connote the brown plastic dial necessary on the leather model. It is probably better to keep things simple and to think of only six models of tube Trans-Oceanics: 7G605 Clipper, 8G005Y, G500, H500, R-520/URR and the 600 Series.

Counting the transistorized models is similarly complex. The simple version of this model sequence is that there were four solid-state Trans-Oceanics: Royal 1000, Royal 3000, Royal 7000 Series and R-7000. However, each model had at least one other version or "sub-model." Again, the simple approach of counting only four solid-state models seems best. Figure 6-1 is a complete listing of the Trans-Oceanic models.

Exactly when each model was produced and sold has become a bit blurred in the radio hobby communities and in the hobby press. This is probably due to the similarity of several of the models and to the fact that Zenith Electronics Corporation and many of its dealers are no longer in the radio business. Figure 6-1 presents data on the years of sale of each model that are both accurate and complete. They are based on press releases (at model introduction), dealer brochures, Zenith catalogs and the print advertising of the era. Exactly when each model left the Zenith line has also become blurred in the hobby communities. This is partly due to the fact that some dealers were actively selling various models for some time after they had "officially" left the Zenith line. Again, the information in Figure 6-1 is drawn directly from Zenith catalogs and from information in the Zenith corporate files. It is definitive.

SIMPLE APPROACH	YEAR(S)*	COMPLETE APPROACH	YEAR(S)*
7G605 CLIPPER	1942	7G605 CLIPPER	1942
8G005Y SERIES	1946-49	8G005Y	1946-47
G500	1949-50	8G005YTZ1	1948-49
H500	1951-53	8G005YTZ2	1949
R-520/URR	1953	G500	1950-51
600 SERIES	1954-62	H500	1951-53
ROYAL 1000 SERIES	1958-67	R-520/URR	1953
ROYAL 3000 SERIES	1963-71	L600	1954-55
ROYAL 7000 SERIES	1969-77	R600	1955
R-7000 SERIES	1979-81	T600	1955
		Y600	1956-57
		A600	1958
		B600	1959-62
10 TOTAL MODELS		ROYAL 1000	1958-63
		ROYAL 1000-D	1959-62
		ROYAL 1000-1	1963-67
		ROYAL 3000	1963
		ROYAL 3000-1	1964-71
		ROYAL 7000	1969-70
		ROYAL 7000Y-1	1971-72
		ROYAL D7000Y	1973-78
		R-7000	1979**
		R-7000-1	1980
		R-7000-2	1981
		24 TOTAL MODELS	

*The dates given are the "model year" of the radio as shown in Zenith catalogs and corporate records. For instance, the Royal 1000 was introduced in the late fall of 1957 for the Christmas season. However, it was part of the 1958 Zenith line.

** Radio enthusiasts in the last decade have come to refer to the R-7000 as the "R-7000 (Taiwan)." This is erroneous. The first year of R-7000 production took place in Chicago. The R-7000-1 (1980) and the R-7000-2 (1981) were each assembled in Taiwan from parts produced in the United States (refer also to endnote 4 for Chapter 6).

Figure 6-1: TRANS-OCEANIC MODELS (*see also the* Endpiece)

Many, but not all, models were introduced in October or November for the Christmas season. They were officially a model of the following year in most cases. Thus, a radio offered for sale in November 1948 was actually a "1949 model."

Number Produced

When Zenith left the radio business in 1982, major portions of the production records were removed to a remote location and are not now accessible. Zenith corporate records do indicate that the one-millionth Trans-Oceanic was produced in August 1964. That would surely have been a Royal 3000-1 model. In all likelihood, the total production for all models of Trans-Oceanics was less than 1,500,000 units. (*see also the* Endpiece)

Spectrum Coverage

As originally conceived, the Trans-Oceanic was a specialized, as opposed to a "general coverage," shortwave receiver. The 7G605 Clipper was designed to cover the medium wave (AM Broadcast) band and the five quite narrow segments of the shortwave spectrum which were dedicated to International Shortwave Broadcasting. In general, the medium wave spectrum runs from .5 MHz to nearly 2.0 MHz. The short-

wave spectrum runs from 2.0 to 30 MHz. The Commander and his Zenith engineers decided to concentrate the shortwave coverage of the Trans-Oceanic on only those five narrow segments of band which actually contained stations broadcasting from overseas. This strategy allowed the engineers to spread each narrow spectrum of frequency across an entire Trans-Oceanic dial line, making it much easier to find and tune in a particular station. The remainder of the shortwave spectrum contained many signals: military, police, commercial communications, amateurs and teletype. Tuning of these signals was largely ignored so that the average listener could more successfully tune in elusive shortwave broadcast signals.

The commitment to shortwave broadcast listening was somewhat modified when the least popular SW broadcast band (49 meters) was dropped and two "continuous coverage" bands were added in the "all new" H500 model. These two continuous coverage bands covered from 2 to 8 MHz. According to advertising of the time, these coverage changes were made to accommodate sportsmen, explorers and yachtsmen needing access to maritime and navigational signals found between 2 and 8 MHz in those days.

The commitment to specialized bands, which spread each International Broadcast Band across a separate dial, was maintained throughout the 40 years of Trans-Oceanic models until the introduction of the ill-fated R-7000, the last Trans-Oceanic. Figure 6-2 illustrates the spectrum coverage of each of the Trans-Oceanic models.

SPECTRUM COVERAGE
ZENITH TRANS-OCEANIC MODELS

MODEL	DATE	LONG WAVE	MEDIUM WAVE	2-4 MHz	4-8 MHz	49 Meters	31 Meters	25 Meters	19 Meters	16 Meters	13 Meters	27 MHz CB	FM	VHF AIRCRAFT	VHF HIGH
7G605	1942		X			X	X	X	X	X	X				
8G005Y	1946		X			X	X	X	X	X	X				
G500	1950		X			X	X	X	X	X	X				
H500	1951		X			X	X	X	X	X	X				
R-520/URR	1953		X			X	X	X	X	X	X				
600 SERIES	1954		X			X	X	X	X	X	X				
ROYAL 1000	1958		X		1	X	X	X	X	X	X				
ROYAL 1000-D	1959	X	X		1	X	X	X	X	X	X				
ROYAL 3000	1963	X	X		1	X	X	X	X	X	X		X		
ROYAL 7000	1969	X	X		2	X	X	X	X	X	X		X	3	
ROYAL 7000Y-1	1971	X	X		2	X	X	X	X	X	X		X	4	
ROYAL D7000Y	1973	X	X		2	X	X	X	X	X	X		X	X	
R-7000	1979	■	■	■	■	■	■	■	■	■	■	■	■	■	■

1 4 to 9 MHz
2 3.5 to 9 MHz
3 single crystal-controlled frequency
4 interchangeable crystal-controlled VHF frequencies

■ R-7000 offers continuous SW coverage from 1.8 MHz to 30 MHz in 6 bands, plus a spread band for the 40 CB channels (26.8 to 27.6 MHz)

Retail Price

The Zenith Trans-Oceanic was a very expensive radio. Throughout the first two-thirds of its forty-year life, it was the most expensive portable radio then offered for sale. The opening retail price of the 7G605 Clipper in late 1941 was $75.[5] Viewed from a half century later, this does not seem to be an exorbitant amount to pay for the only portable receiver in the world which could tune in broadcasts from overseas. However, if we properly factor in a half century of inflation, the retail price of $75 for the Trans-Oceanic Clipper was actually $695 in 1994 dollars!

Figure 6-3 is a graph which depicts both the actual retail price for each model in each year and the "Adjusted Retail Price" with the correct inflation factor taken into account each year. It is most instructive to study and compare these two curves.

First, it is rather startling to note that there was very little increase in the actual retail price during the life of any given model. The 8G005Y entered the market in 1946 at $125 and maintained that price throughout its almost four-year life span. The $140 price of the L600 was maintained throughout the eight-year life of the 600 Series. This same level pricing policy held throughout the solid-state years except for the Royal 7000 in the 1970s. Inflation in the 1970s rose to a savage post-war peak, with the dollar losing more than half its value in ten years. The retail price of the Royal 7000 increased by less than 10% during that period.

The effect of inflation on this steady-state retail price and the very high retail price as expressed in 1994 dollars of any Trans-Oceanic is readily apparent from the upper curve. This curve represents the adjusted purchase price of a Trans-Oceanic expressed in 1994 U.S. dollars. These figures were derived by the Department of Economics at Okla-

homa State University and are based on the Consumer Price Index for the years in question. One can see that the tube models varied some in price; those costs clustered, however, around $750 (1994 dollars) throughout their twenty-year life.

The most startling aspect of the entire graph is the near doubling of price between the tube models and their transistorized replacements. The Royal 1000-D, introduced in late 1958, cost a whopping $1,432 in 1994 dollars! From a portable radio user's point of view, the Royal 1000 was vastly superior to all other portables previously offered to the public. Weight had been reduced significantly and the cost of batteries (for the first time cheap "D" flash light cells) was also significantly reduced. Further, the now inexpensive batteries lasted much longer in the new transistorized models. The Commander and his engineers had produced a breakthrough consumer product, they charged an appropriate price, and the market beat a path to local Zenith dealers.

The second startling feature of the "adjusted price" curve is the long and steep slide of adjusted price from the mid-1960s until the closeout of the line. This must have been particularly difficult for Zenith during the Royal 7000 years. Their labor costs must have escalated rapidly during the inflation-ridden 1970s as Zenith continued to pay ever-higher salaries to its workers. It was, of course, impossible to reduce the labor involved in the Royal 7000 in midstream because Zenith found itself trapped by the decision (in 1968) to use the old labor-intensive point-to-point wiring. The Royal 7000 may have been produced at a profit at the beginning of its life; it must surely have been sold at a significant loss by the end.

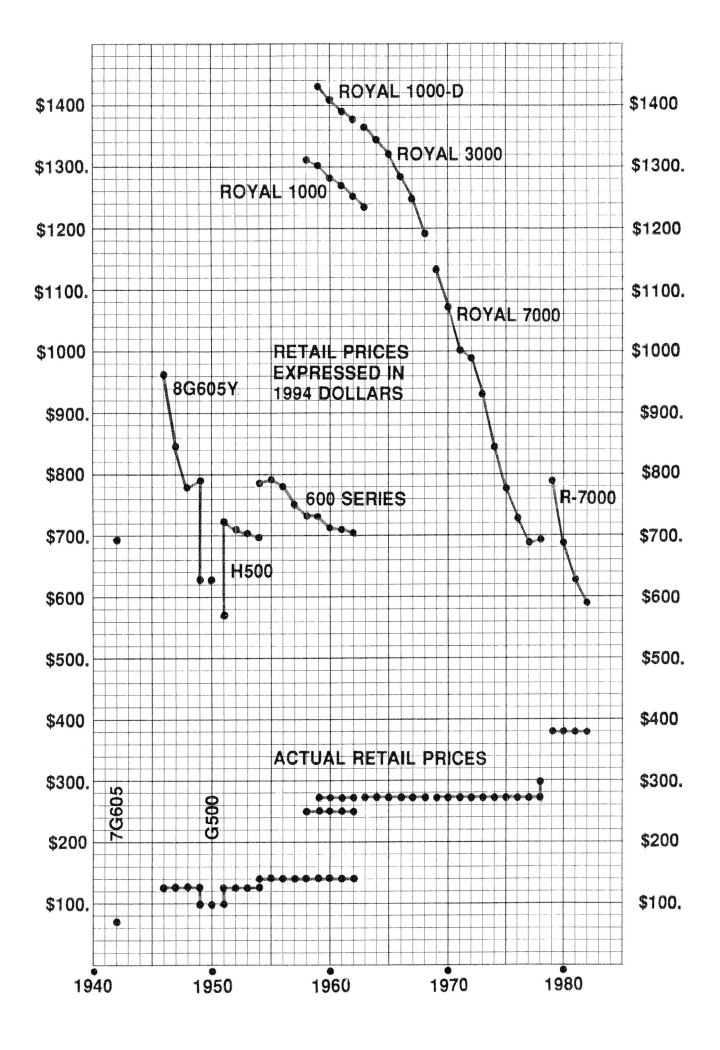

Chapter 7
TRANS-OCEANIC PORTRAITS

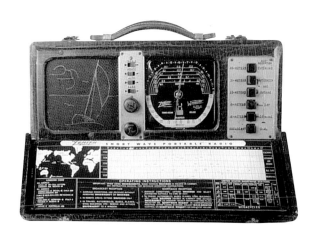

7G605

H500

8G005Y

R-520/URR

G500

600 Series

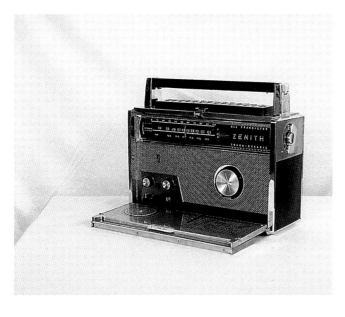

Royal 1000 Series

Royal 7000 Series

Royal 3000 Series

R-7000 Series

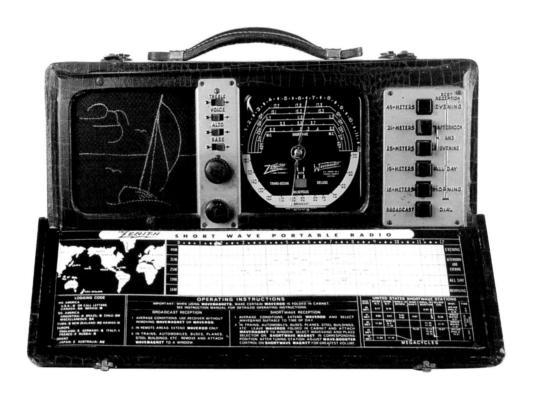

Trans-Oceanic Clipper Model 7G605.

Trans-Oceanic Clipper Model 7G605.

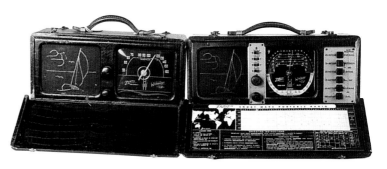

Universal Model 6G601 and Trans-Oceanic Clipper Model 7G605.

Trans-Oceanic Clipper Model 7G605.

7G605
The Trans-Oceanic Clipper

Introduction

After two and one-half years of design and prototype testing, "The Zenith Trans-Oceanic Clipper Shortwave and Broadcast Deluxe Portable Radio" was introduced in January 1942, and it was an instant success as a portable for use in receiving broadcasts carrying news of World War II. The Clipper was used as a privately purchased portable by military personnel abroad as well as a reasonably priced receiver for civilians at home.

Among the notes from the Zenith archives are: "Admiral McMillan put prototype into Arctic service in 1940" (from back of a photograph of a 7G605); "100,000 orders in hand when Zenith went over to 100% Wartime production in April 1942, many produced before that time" (before shutdown for war production).

Despite the "100,000" outstanding orders, about a thousand of the final production run of the Model 7G605 Clipper were withheld from sales by Commander McDonald. Throughout the war years, these were presented to lucky senior military officers and to various other dignitaries. A photograph in the Zenith archives shows a very young Frank Sinatra receiving a Clipper.

Versions

There were two versions of the 7G605 Clipper: the original version had a sailboat embroidered with white thread on the black speaker grille cloth; the second, and most common, version had an unidentified four-engine bomber, similar to a Boeing B-17, embroidered on the grille cloth. Commander McDonald directed the change to the "bomber grille" in a December 17, 1942, memo but cautioned that the change should only be made after the existing sailboat grille cloths ran out (memo in the Zenith archives). The sailboat grille was the design shown in the owner's manual and in early photographs. Advertising (January and April 1942), and other photographs clearly taken after the U.S. entered World War II, show the "bomber version." Bombers were used on most, if not all, of the units of both the second and third production run. In both versions, the 7G605 was called the "Zenith Trans-Oceanic Clipper."

Number Produced

There were 35,000 7G605 Trans-Oceanic Clippers produced in runs of 10,000, 20,000 and 5,000, in that order. The production line was shut down in April 1942 for full-time war production and a ceremony was held with Commander McDonald soldering the last connection and running tests on the last Trans-Oceanic. Many photographs of this ceremony were found in the Zenith archives. The serial numbers and production numbers were:

START	END	CHASSIS	NUMBER PRODUCED
T-847011	T-857010	7B04	10,000
T-860261	T-880560	7B04	20,000
T-885261	T-890260	7B04	5,000

LINE SHUT DOWN, T-SERIES SERIAL NUMBERS END

Chassis: 7B04.

Cabinet Design

The basic design and proportions of the cabinet, including its formed leather handle and hinge hardware, are the same as those found on the 1942 Model 6G601 Universal, which was introduced four months earlier. The imitation alligator hide covering is unique to the 1942 Trans-Oceanic and companion Universal. The 1942 Universal also came in two colors of airplane fabric, as well as cowhide, as had previous Universal models.

The speaker grille of both the Trans-Oceanic and the 6G601 Universal is the most prominent feature of these unique radios. All Universal speaker grilles were black with the sailboat embroidered in white thread, the same design as the sailboat version of the 7G605; the Trans-Oceanic was produced with either the sailboat or the four-engine bomber.

The two metal escutcheons on the front panel and the Waverod are plated with nickel. The dial face of the Trans-Oceanic is unique, but it is closely related to the Universal dial and to the dials of other small Zenith radios of 1941 and 1942.

Electronic Design

The electronic design of the 7G605 is both derivative and highly innovative. After the Commander's famous message from the *Mizpah* in Lake Huron dictating the research and development effort, Zenith project engineers began their design with one of the Universal portables, the 1940 Model 5G401, taken directly from the assembly line. This receiver was based on "loktal" vacuum tubes, which snapped firmly in their sockets, rather than the simple friction fit of the normal tubes of the day. The loktal tubes had been developed in the late 1930s for military and other rugged applications.

Two major technical hurdles stood between the Universal circuitry and a successful Trans-Oceanic. First, the receiver had to be made more sensitive, since shortwave radio broadcasts from overseas produced considerably weaker signals than were found on the more local AM band. This was solved by some redesign of the circuitry. The much more difficult task was getting the oscillator tube to work predictably (or at all) at shortwave frequencies. In 1941, employing the relatively new 1.5-volt tubes necessary in a battery-powered portable, this was nearly impossible. Most other radios used at least 6-volt tubes. Once the engineers were able to get shortwave frequency oscillations from a 1.5-volt tube, the challenge was to do so at a stable frequency, whether on weak or strong batteries or on the much less expensive 117-volt AC wall current.

The other major electronic design decision was the commitment to the multi-band, bandspread dial. This type dial, focusing as it did on the International Shortwave Broadcast Bands, had been used previously on only one or two of the most expensive American console models. Commander McDonald himself made this design commitment after becoming frustrated at having to employ "micrometric" fingers to tune an early Trans-Oceanic prototype with a conventionally crowded and touchy shortwave dial.

Designing the first Trans-Oceanic was quite a struggle and was probably the reason that the Clipper was not shown in the 1942 Zenith dealer brochures. When all was said and done, however, the Commander and his Zenith engineers had created the world's first battery-powered portable all-wave radio. It was introduced about two months after the rest of the 1942 line, just in time to take its place in radio history.

Commander McDonald's close friend, Powell Crosley, Jr., President of Crosley Radio, was *very* impressed with the Trans-Oceanic. Hearing of Crosley's regard for the Clipper, the Commander had a special model 7G605 made for his friend. The Zenith plant cast a customized dial pointer, replacing the Zenith lightning-bolt "Z" encircled with a brass ring with an encircled "C," thus creating the first, and last, CROSLEY Trans-Oceanic.[1]

Last Offered

Production of the 7G605 was halted on April 22, 1942, with a unit carrying the serial number T-890260.

Price

Some hobby and scholarly sources list the retail price of the 7G605 as $75 while others indicate it at $100; the correct retail price was $75. Zenith corporate records and the Commander's private files both are very clear that the official retail price of the 7G605 was $75 throughout its short retail life.

Companion

The natural companion to the 7G605 Clipper was the "Universal" 6G601 MW-only portable with the same type case and sailboat speaker grille. The earlier Universals in 1940 and 1941 (Models 5G401 and 6G501) had introduced the detachable Wavemagnet, and were very similar to the 1942 Universal Model 6G601. The earlier Universals did not, however, feature the sailboat on the speaker grille, nor did they come in the simulated alligator hide-covered cabinet of the 6G601 Universal and the 7G605 Trans-Oceanic.

Selected Print Advertising

Since the 7G605 Clipper was introduced at almost the exact moment of the American entry to World War II, the print advertising campaign for the new radio was very brief. Print advertisements of the 7G605 were published in the January and April 1942 issues of *National Geographic* and in the following issues of the *Saturday Evening Post*: January 24, 1942, page 55; March 14, 1942, page 45; May 16, 1942, page 62; and, June 20, 1942, page 54. Many patriotic wartime Zenith advertisements also featured the 7G605; they were not published to market the radio, however, since none were being produced. This advertising strategy, and the great success of the 7G605 at home and with the troops in the field, built an almost impossibly high demand for post-war Trans-Oceanics.

SENSATIONAL NEW TRANS-OCEANIC PORTABLE RADIO

ONLY *Zenith* HAS THIS!

LIMITS OF PORTABLE RADIO LISTENING RANGE NOW REMOVED

The first and only portable radio guaranteed to receive Europe, South America or the Orient—every day, even on planes, trains and ships—or your money back. *This guarantee has been made possible by the new Shortwavemagnet.*

Never Before Offered to the General Public

NEW SUPER-DISTANCE...
SHORT-WAVE AND BROADCAST
DE LUXE PORTABLE RADIO

FIRST TIME! Personal short-wave radio reception from our own or foreign continents—while you ride in planes, trains or ships!

FIRST TIME! A portable radio that gives domestic short-wave reception in locations where broadcast does not penetrate in the daytime.

FIRST TIME! The miraculous time and band buttons. Pre-set the pointer—"Press a button . . . there's Berlin!"

FIRST TIME! On conveyances—on land—sea—air—choice of portable radio reception with built-in movable broadcast and short Wave-magnets.

FIRST TIME! Band Spread makes foreign station tuning on a portable radio as easy and simple as ordinary radio broadcast tuning.

FIRST TIME! Logged at the factory on short wave broadcasts...A convenient logging chart on inside lid of cover is pre-logged by factory experts. Shows exactly what stations are found on each wave band and at what number on the dial.

FIRST TIME! Zenith Famous Radiorgan Tone Device on a portable radio.

RESULTS—In test air flights, train trips and cruises, have amazed even the country's best radio research brains.

TESTS—In far-away vacation, fishing and hunting locations where daytime radio *broadcast* simply is not heard—showed *short-wave* radio bringing in favorite programs direct from New York, Boston, Schenectady, etc., over the super-distance air waves.

POWER—From self-contained battery and standard lighting current ingeniously interchangeable at a second's notice. Also, Telescope whip aerial for use in getting *extra* distance.

COMING—In a week or two. Watch your Zenith dealer's window. Or better leave your name and address so he may telephone you—Don't miss this NEW ONE!

ANOTHER GREAT ZENITH FIRST

ZENITH

TRANS-OCEAN CLIPPER
DE LUXE PORTABLE RADIO
With the Built-in Movable Wavemagnets
U. S. Patent No. 2164251

ZENITH LONG DISTANCE RADIO...ALWAYS A YEAR AHEAD

53

ZENITH

REG. U.S. PAT. OFF.

TRANS-OCEAN CLIPPER DE LUXE STANDARD AND SHORTWAVE PORTABLE RADIO

WAVEROD OPERATION
(TELESCOPE AERIAL)

. . . for super-efficient short-wave reception when not confined within steel-shielded construction. Telescope type that extends to 5 ft. Automatically switched off when folded into case.

WAVEMAGNET AND SHORTWAVE MAGNET OPERATION

. . . for reception on ships, planes, trains, autos or in steel-shielded buildings. (U.S. Patent No. 2,164,251)

STANDARD OPERATION

. . . with the Waverod and Wavemagnets left in case, as shown, for average listening under ordinary conditions to standard broadcasts.

ONLY ZENITH HAS ALL THESE TESTED SELLING FEATURES

SHORTWAVE MAGNET: The secret of short wave reception on land, sea or in the air . . . where other portables fail.

WAVEMAGNET: Built-in movable for efficient standard broadcast reception when in planes, trains, autos or in windowed steel shielded buildings.

WAVE BOOSTER: For super-sensitive shortwave reception. Aids tuning.

WAVEROD: For peak efficiency in shortwave reception, when not confined in steel-shielded construction.

TIME BAND SELECTOR: Eliminates guess work for efficient short wave reception. Simply press the button designating the time of day and dial to the waveband indicated.

RADIORGAN: Your choice of 16 tone combinations in four button controls. Now, this famous Zenith feature is yours in a portable.

MAGNASCOPE SPREAD BAND TUNING: Enables you to define and separate strong foreign stations which tune broader than ordinary broadcast stations.

FIELD TESTED: On planes, trains, and ships—throughout America, in the tropics . . . and in the Arctic.

LOGGING CHART: Now, for the first time, a portable shortwave radio pre-logged at the factory for all principal short wave stations.

SIX WAVE BANDS: Full shortwave and broadcast coverage to allow maximum reception at all times of day or night.

Specifications

TUBES: Seven tube superheterodyne, including rectifier.

SIX WAVE BANDS: Standard broadcasts, 540-1620 Kc.; 49 meter band, 6.0-6.5 Mc.; 31 meter band, 9.4-9.8 Mc.; 25 meter band, 11.7-11.9 Mc.; 19 meter band, 15.1-15.3 Mc.; 16 meter band, 17.6-18.0 Mc.

THREE-WAY POWER OPERATION: Operates from Zenith Z-985 battery pack and two flashlight cells or 110 volt AC or DC.

CABINET: Smart luggage styling with deluxe hardware . . . simulated brown alligator covering. Removable front cover. Overall dimensions: Height, 11⅛ in., Width, 16⅝ in., Depth, 7½ in.

CONDENSER: Three-gang.

FAMOUS ZENITH FEATURES: Safety Switch, Triple Hi-Ficiency Switch, Guardian Reminder, P. M. Dynamic Speaker, On-and-Off Indicator, etc.

THE FIRST AND ONLY PORTABLE RADIO GUARANTEED TO RECEIVE EUROPE, SOUTH AMERICA OR THE ORIENT EVERY DAY—OR YOUR MONEY BACK. IT ALSO BRINGS IN FOREIGN SHORTWAVE RECEPTION ON TRAINS, PLANES AND SHIPS.

SAKS FIFTH AVENUE — NEW YORK CITY

BIRKEL-RICHARDSON MUSIC CO.
LOS ANGELES, CALIFORNIA

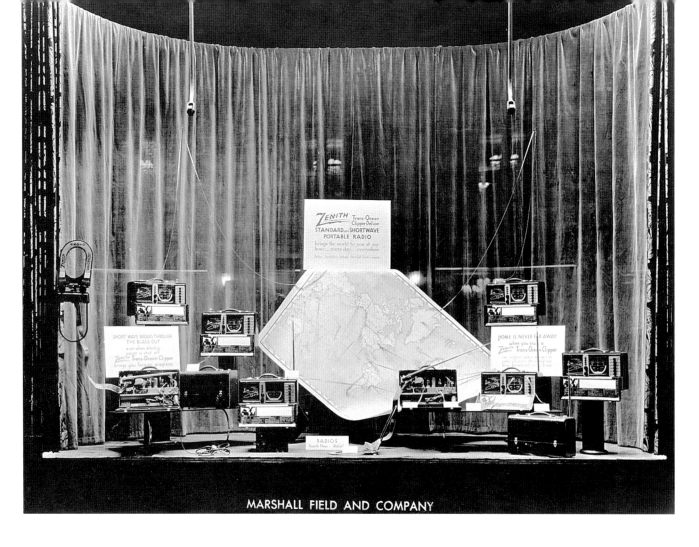

MARSHALL FIELD AND COMPANY

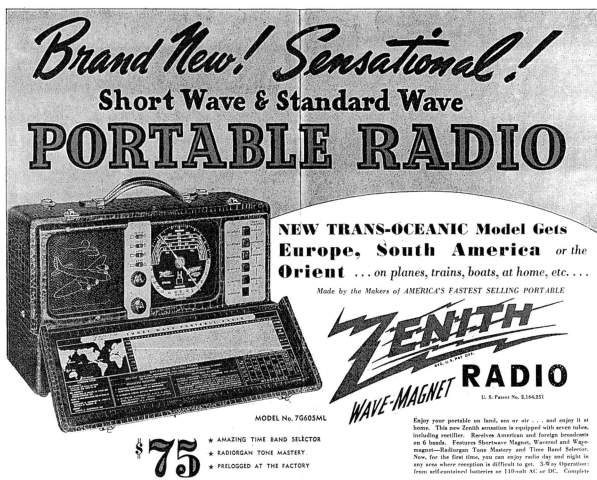

Brand New! Sensational!

Short Wave & Standard Wave
PORTABLE RADIO

NEW TRANS-OCEANIC Model Gets
Europe, South America *or the*
Orient *... on planes, trains, boats, at home, etc. ...*

Made by the Makers of AMERICA'S FASTEST SELLING PORTABLE

Zenith

WAVE-MAGNET **RADIO**

U. S. Patent No. 2,164,251

MODEL No. 7G605ML

$**75**

★ AMAZING TIME BAND SELECTOR
★ RADIORGAN TONE MASTERY
★ PRELOGGED AT THE FACTORY

Enjoy your portable on land, sea or air . . . and enjoy it at home. This new Zenith sensation is equipped with seven tubes, including rectifier. Receives American and foreign broadcasts on 6 bands. Features Shortwave Magnet, Waverod and Wave-magnet—Radiorgan Tone Mastery and Time Band Selector. Now, for the first time, you can enjoy radio day and night in any area where reception is difficult to get. 3-Way Operation: from self-contained batteries or 110-volt AC or DC. Complete

56

Universal Model 6G601.

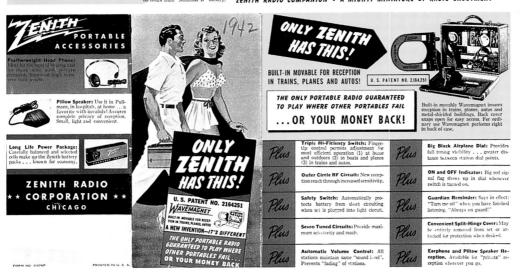

6G601
Universal
Companion to the 7G605
Trans-Oceanic Clipper

Introduction

Unlike its companion 7G605 Trans-Oceanic, the 6G601 Universal was introduced with the rest of the Zenith 1942 line in the late fall of 1941 and appeared in the standard 1942 Zenith dealers' brochure. Studies of the serial numbers indicate that production of the Universal began considerably earlier than did that of the Trans-Oceanic. It is conjecture, but logical to assume, that technical development of the new Trans-Oceanic was lagging, and it was decided to delay its production beyond the rest of the 1942 line.

Versions

There were four versions of the 6G601. The only difference between them was the covering of the cabinet:

6G601D	Blue and Gray Airplane Fabric
6G601MH	Brown and Ivory Airplane Fabric
6G601L	Genuine Cowhide
6G601ML	Simulated Brown Alligator

Since the 7G605 Trans-Oceanic was available only in Brown Alligator, purists would consider only the latter version Universal, the 6G601ML in Brown Alligator, as the companion to the Trans-Oceanic. All known 6G601s exhibit the sailboat embroidered on the black speaker grille.

Chassis

All versions were produced with the 6B03 chassis.

Last Offered

Production was closed-out in April 1942 with a 5000 unit "rump" run made immediately before the last Trans-Oceanic run.

Number Produced

There were over 225,000 1942 Universal portables produced. The serial numbers and production runs were:

START	END	NUMBER PRODUCED
T-527751	T-530250	2,500
T-533301	T-562400	29,100
T-606611	T-606860	250
T-606861	T-681360	74,500
T-715711	T-778710	63,000
T-794211	T-844210	50,000
T-880261	T-885260	5,000

TOTAL PRODUCED: 225,350 units

Design

Both the 6G601 Universal and its companion 7G605 Trans-Oceanic have a common ancestor: the 1940 Zenith Universal portable. That 1940 Universal, Model 5G401, was the first example of the Zenith Universal marque; it also introduced Commander McDonald's own invention, the newly developed and patented Wavemagnet. There were two other "suitcase" Zenith portables in the 1940 Line, each with four tubes.

The 1941 line included at least four portables: the top-of-the-line 6G501 Universal (with a sixth tube) followed by two lesser portables and the new Zenith "Poket Radio." The Universal cabinet was continued as a two-door (front and back) suitcase and was available in either "Airplane Fabric" or leather.

The 1942 Universal 6G601 was the culmination of several years of development and was one of the best AM portables built using pre-war technology. According to advertising copy, it used two dual-function tubes, giving it the power of an eight-tube radio.

The case of the 6G601 was larger than the earlier Universals and had a more rounded side profile. One has the sense that this, finally, was a mature product. For the first time, the appearance of the front panel was more than just utilitarian: the sloop-rigged sailboat was the first attempt on any Zenith portable to bring beauty into the appearance of the radio. Some evidence suggests that Robert Davol Budlong *may* have had a hand in the design of the 6G601 (but probably not its companion Trans-Oceanic Clipper).

Price

$29.95, including battery.

Selected Advertising

The Universals were widely advertised in 1940 and 1941, primarily prior to the summer season. The 1942 Universal was not advertised extensively, due to the changeover to wartime production in the spring of 1942.

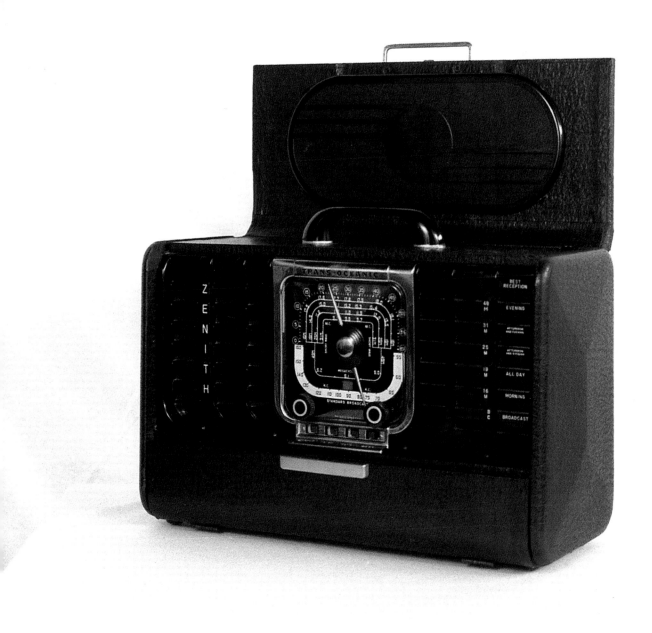

Trans-Oceanic Model 8G005Y.

Trans-Oceanic Model 8G005Y.

Universal Model 6G001Y and Trans-Oceanic Model 8G005Y.

Trans-Oceanic Model 8G005Y.

The 8G005Y Series
The Post-War Trans-Oceanic

Introduction

The post-war Trans-Oceanic was "teased" with a non-specific "Back Soon-Better than Ever" magazine ad in November 1945. The 8G005Y was actually introduced in April 1946, with a massive spring ad blitz in major national magazines. A typical advertisement was a full page ad in *Better Homes and Gardens*, May 1946, with the headline: "NEW SUPER POWER TRANS-OCEANIC PORTABLE RADIO." These early advertisements also introduced "companion radios": the new Zenith Global and Zenith Universal Portable.

Versions and Chassis

8G005Y (chassis 8C40)	1946 and 1947 Lines
8G005YTZ1 (chassis 8C40TZ1)	1948 Line
8G005YTZ2 (chassis 8C40TZ2)	1949 Line

The TZ1 and TZ2 variants denote upgrades of the original AC power supply. The 8C40 chassis used a loktal tube rectifier, while both the TZ1 and TZ2 variants used the much more modern all-glass miniature tube rectifiers.

Last Offered

In the 1949 Line.

Cabinet Design

The design of the all-new 8G005Y cabinet by Budlong's firm set the direction for the design of all tube-model Trans-Oceanics for the next 17 years. Indeed, the upwardly rotating front door introduced by industrial designer Budlong remained a "signature feature" of Zenith Trans-Oceanics from 1946 until the demise of the line in 1982. The cardboard AM Wavemagnet of the 7G605 was replaced by a capsule-shaped plastic unit clipped to the inner side of the front door, while the metal D-shaped Shortwave Wavemagnet was retained throughout the 8G005Y Series.

Materials

The extensive use of war-developed plastics in the 8G005Y was the precursor of adoption of these new materials throughout the radio industry. The "Black Stag" cabinet covering was one of the first uses of what would become known as "vinyl-covered fabrics." This tough covering was a major improvement over the false "alligator-hide" or "airplane cloth" of late 1930s luggage and portable radios.

Electronic Design

Bringing out a pre-war radio in a post-war cabinet was a strategy which most manufacturers used to meet the immediate post-war surge of demand in the civilian radio market. It was generally not possible to design new receivers, test them and get the production lines rolling in the brief period between the end of World War II and the impact of pent-up consumer demand for consumer radios. Although the 8G005Y and its pre-war ancestor, the 7G605 Clipper, share the basic super heterodyne circuit design and even have somewhat similar tube complements, the 8G005Y cannot be considered a pre-war radio in a post-war cabinet.

While maintaining the pre-war loktal-type tubes, the 8G005Y circuit design was a more modern design with several times the number of components dedicated to improvements of selectivity, sensitivity and audio quality than was the case in its ancestor, Model 7G605. The 8G005Y was a much better radio with more components deployed in a more sophisticated manner. It was also the most electronically complex tube-type Trans-Oceanic ever produced.

Price

The price at introduction was $120. By February 1947, it had risen to $124.40, followed by an increase to $128.40 by February 1948. That price held until October 1949, when it was reduced to $99.95. This latter figure was probably a "clearance price" to move inventory before the upcoming introduction of the G500 in December 1949, although it was not mentioned as such in the advertisements.

Companion

The companion receivers at the introduction of the 8G005Y were the Universal Model 6G001Y (from 1946 through the 1947 line - $54.60) and the rare Global Model 6G004Y, offered in only the 1946 line. A new and rather ugly all plastic Universal Model 6E40 (rare) was introduced in the 1948 line and sold for $54.70. It was referred to as the "Batwing Door" model and was *not* a companion to the Trans-Oceanics.

Selected Print Advertising

COLLIERS	Nov 3, 1945, p.81, "Back Soon, Better Than Ever"	
COLLIERS	May 4, 1946, p.27, Introduction Advertisement	
SAT EVE	May 4, 1946, p.121, Introduction Advertisement	
COLLIERS	Nov 30, 1946, p.61, "Under the Christmas Tree" color	
HOLIDAY	Oct 1946, p.64, "Hunting Lodge"	
NAT GEO	Feb 1947, F.A.,"Beach Scene" also *Holiday*	
NAT GEO	Apr 1947, F.A., "Around the Campfire" also *Holiday*	
NAT GEO	June 1947, B.A., "Z Portables Won't Play...Underwater"	
HOLIDAY	Feb 1948, p. 9, "Aristocratic Poolside Scene"	

Happier Holidays for You...

...WITH THE WORLD'S SMARTEST,
FINEST PERFORMING PORTABLE RADIO!

COPYRIGHT 1947, ZENITH RADIO CORPORATION

Wherever you go, wherever you are...North, South, East, West...in the mountains or on the beach ...hunting or golfing...just loafing or (if you're ambitious) exploring the Arctic or Antarctic...the new Zenith Trans-Oceanic Portable puts a *world* of radio at your fingertips! For it's the new, even finer, more powerful model of the sensational radio that kept GI's in touch with home—from steaming jungles and frozen waste lands. It spans continents and oceans just as well in trains, planes, boats, motor cars, remote areas, as it does in your own living room or den! To tune in any of the five international short-wave bands, simply press a button and dial them as easily as local stations. All this—yet you carry the Zenith Trans-Oceanic as easily as a piece of luggage! Today—ask your dealer for a thrilling demonstration.

* * *

Operates on long-life battery pack (up to 1 year normal usage) as well as on AC or DC current. Specially treated for high humidity...works on land or sea, in the tropics, where untreated radios fail. Model 8G005Y (7 radio tubes plus power rectifier) $114.40, less battery. Other Zenith Radios from $26.95 to $395 (West Coast prices slightly higher).

NEW  TRANS-OCEANIC PORTABLE

ALSO MAKERS OF ZENITH RADIONIC HEARING AID

ZENITH RADIO CORPORATION, CHICAGO 39, ILLINOIS • 30 YEARS OF "KNOW-HOW" IN RADIONICS EXCLUSIVELY

65

GIVE A WORLD OF HAPPINESS...

with the World's Smartest, Most Powerful Portable Radio

The whole family will shout for joy when they see their
ZENITH TRANS-OCEANIC "CLIPPER" sparkling beneath the Christmas tree.
For here is *more* than just a gift. Here is lasting pleasure and happiness day after day to
enjoy wherever you live and wherever you go. The "CLIPPER" is equally at home in
living room or faraway vacation spot. So powerful it plays where others won't
—in trains, planes, motor cars, boats, steel buildings, remote areas. And to get any
of the 5 international short wave bands—just press a button and dial them in as easily
as you do local stations. Yes—there's a world of happiness in a ZENITH "CLIPPER"
—for the whole family and especially for the "Santa Claus" who gives it.

*Operates on AC or DC current, as well as long-life battery pack (up to 2 years
normal usage). No bothersome re-charging, no acid or wet battery mess. Model
8G005Y (7 radio tubes, plus power rectifier) styled like finest luggage.*

NEW Zenith RADIO TRANS-OCEANIC CLIPPER

•LONG DISTANCE•
ZENITH RADIO CORPORATION, CHICAGO 39, ILLINOIS

COPYRIGHT 1946, ZENITH RADIO CORP.

Universal Model 6G001Y.

6G001Y
6G004Y
Universal and Global Portables
Companions to the 8G005Y Trans-Oceanic

Introduction

Both the 6G001Y Universal and the 6G004Y Global portables were introduced at the same time as the post-war 8G005Y Trans-Oceanic, in April 1946. The Global appeared in only the earliest print advertisements in the spring of 1946. Such a sparse presence in advertising has led to speculation that the 6G004 was never taken beyond the prototype stage. However, the Global was produced for the 1946 line, though in very low numbers. It is exceedingly rare today.

Versions

6G001Y
This is the first and most common version. It utilized all loktal tubes, including the rectifier, 117Z6.

6G001YZX
This is a very rare version of the 6G001Y, with the cabinet made entirely from *aluminum* formed to emulate the wooden cabinet used on all other versions. It is entirely speculation, but this unusual model (and its 'X' designation) was probably an experiment to determine the economics of shifting from wood to aluminum cabinets on Zenith portables. Commander McDonald wrote several internal memoranda, even before World War II, urging Zenith engineers to experiment with aluminum, then considered a marvelous new material. *NOTE*: This "YZX" version contained the original 6G001 all loktal tube complement.

6G001YTZ1
Like its companion Trans-Oceanic 8G005Y, later runs of the 6G001Y Universal used a miniature tube rather than a loktal tube-based power supply. This is the only difference between the 6G001Y and the 6G001YTZ1 versions. There may also have been a few aluminum cabineted 6G001YTZ1Xs assembled; if this is true, it would date the aluminum cabinet experiments to sometime in 1947.

GLOBAL 6G004Y
As far as is known, there was only one version (all loktal-tube) of this now very rare radio.

Chassis

6G001 Universal -- 6C40
6G004 Global -- 6C41

Last Offered

The 6G001 Universal was displaced in the Zenith 1948 Line by the unusual 6G801 "Batwing Door" Universal model. The 6G004 Global was not produced after 1946.

Design

Electronically, the 6G001 Universal was one of the finest portable AM radios offered in the post-war market. Its circuits employed six tubes rather than the four or five tubes then common in portables and small table models. It was, and is, an excellent AM radio producing very mellow audio, thanks to its comparatively large cabinet/sound box.

Nothing is known about the performance of the Global other than it offered coverage of shortwave (9.4 to 12.1 MHz) in addition to the AM broadcast band.

With the 6G001/6G004 design, Robert Davol Budlong and his staff produced one of the most beautiful portable radios of the tube era. The radio was designed as a companion to the 8G005Y Trans-Oceanic, but also related aesthetically to several of the more popular Budlong designs for Zenith table radios (e.g., 6D015 and similar). The slender proportions of the cabinet and the elegantly laid out front panel are particularly admired today.

Price

6G001Y Universal: $54.60 in mid-1947.
6G004Y Global: Unknown (about $80?).

Companion

Both the Universal 6G001Y and the Global 6G004Y were designed and advertised as companions to the 8G005Y Trans-Oceanic.

Selected Print Advertising

SAT EVE	Apr 6, 1946, p.89, (Universal only)
SAT EVE	May 5, 1946, p.121, (Universal and Global)
BET HOM	May 1946, p.13, (Universal and Global)
HOLIDAY	Oct 1946, p.64, (Universal)
NAT GEO	June 1947, B.A., (Universal)
HOLIDAY	Oct 1947, p.104, (Last advertisement for the 6G001Y Universal)

Trans-Oceanic Model G500.

Trans-Oceanic Model G500.

Universal Model G503 and Trans-Oceanic Model G500.

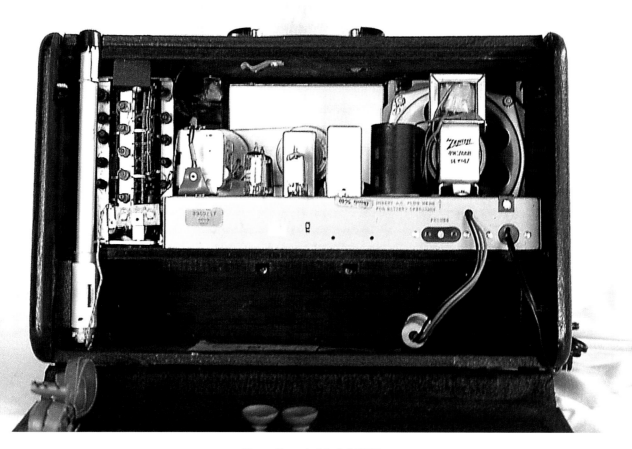

Trans-Oceanic Model G500.

G500
The Changeover Model

Introduction

By 1948, the 8G005Y was technologically obsolete: except for the rectifier, it still used the large, costly and comparatively inefficient loktal tubes developed in the late 1930s. Zenith engineers were charged with designing a state of the art chassis based on the new all-glass miniature vacuum tubes and a solid-state rectifier. The result was an electronic design which formed the basic platform for all remaining tube model Trans-Oceanics. New cabinet and front face designs were not available when the new 5G40 chassis was completed, however, so Budlong's firm "spruced up" the cabinet of the 8G005Y to temporarily house the new chassis. Thus was born the G-500 "changeover" model. It was introduced in October/November 1949 for the 1950 Line. The G500 remained in production for only 18 months until the "all-new" H500 succeeded it. This very short life, and thus low production numbers, makes the G500 the rarest of the post-war tube-type Trans-Oceanics.

Versions

There was only one version of the G500 Trans-Oceanic and only one chassis, 5G40.

Last Offered

May 1951.

Cabinet Design

There were no design innovations in the exterior appearance of the G500; however, it is a superb example of the development and refinement of the basic appearance of the 8G005Y. The designers and Commander McDonald must have felt that the 8G005Y was a bit too black: the most striking visual element of the G500 is a circular bright brass rendering of the Zenith corporate seal in the center of the black plastic Wavemagnet. Budlong's design for the G500 also introduced the brass hinge/plastic handle design that would continue with only minor changes throughout the tube years. Finally, probably for economic reasons, the cabinet was simplified by eliminating the pull-down log-book door and the subtle beveling of the cabinet sides of the 8G005Y. To many collectors, the G500 is the most beautiful Trans-Oceanic, offering the perfect balance of dark plastic, brass and Black Stag covering; it is a truly beautiful product of early post-war industrial design.

Electronic Design

The design and production of the 1950 5G40 chassis, based on miniature tubes, was *the* major electronic innovation of the Trans-Oceanic series prior to the adoption of the transistor in late 1957. Further, the creative engineering design of the G500 more fully exploited the possibilities of the sophisticated miniature tubes, creating a better, longer playing radio (less battery drain) with only five tubes rather than eight. The audio power amplifier section of the 8G005Y required three tubes (two amplifiers and a phase inverter for push-pull). These tubes were replaced in the G500 by a single 3V4 tube. The fragile and "cranky" 117Z6 rectifier tube of the older set was replaced by a new solid-state selenium rectifier. Thus, the eight-tube 8G005Y was succeeded by a better radio with lower battery drain and three fewer tubes. These cost reductions were coupled with amplifications of the cabinet to allow Zenith to reduce the price of the G500 from the 8G005Y level of $128.40 to $99.50, a reduction of 20% for a more modern radio!

The spectrum coverage, and thus the "RF head" (the tower of band switches and fragile coils to the left of the main chassis) of the G500, was identical to the older 8G005Y receiver since the dial face, plastic front panel and push buttons were not changed from the older models for the short life of "the changeover model." The 5G40 chassis also introduced the plug-in "Power Supply Adapters" accessory to allow operation on the 220-volt AC or DC power often found abroad.

Price

The introductory price of the new G500 was identical to that of the recently reduced 8G005Y that it was replacing: $99.95. That low price held throughout 1950 and apparently through model close-out in mid-1951. This was, by far, the lowest priced Trans-Oceanic.

Companion

A new Universal, Model G503 (chassis 5G41) "Flip Dial Universal," was introduced at $49.95 in March 1950.

Selected Print Advertising

NAT GEO	Aug 1950, F.A., "Outpulls - Outplays - Outperforms Them All!"
HOLIDAY	Aug 1950, p. 60, "Only Zenith" features 4 best Zenith portables

The King of the Season
Recommends
The Royalty of Radio

What wise and sensible givers—these smart people who give more than just gifts. When *you* give Zenith†—the Royalty of Radio—you give years of lasting pleasure. Matchless radio masterpieces of unequalled quality and beauty. Treasured lifetime possessions from the world's leader in portables. And remember too—to *give* a Zenith ... portable, table radio or console ... is to *know* the supreme pleasure of Christmas giving.

New Zenith "TRANS-OCEANIC"

New edition of the world-famous portable that out-performs any other, any time, anywhere! More powerful, more sensitive, yet lighter, easier to carry. Plays in boats, trains, planes, remote areas. Standard Broadcast, plus international Short Wave on 5 separate bands. "Tropic-Treated" against humidity. Works on thrifty, long-life battery, and on AC or DC.

only $99⁹⁵* *Less Batteries*

New "ZENETTE"† by Zenith

Tiny and exquisite as a jewel, yet a giant in power and volume! Has the biggest speaker Zenith has ever used in a set this size. Lift lid, set's playing—close lid, set's off! Plays on battery, AC or DC. Weighs a mere 5½ lbs. Ultra-smart plastic case in maroon, jet black or white.

only $39⁹⁵*
†®   *Less Batteries*

Outpulls—Outplays—Outperforms them all!

NEW ZENITH "Trans-Oceanic"
Standard and Short-wave
PORTABLE RADIO

"Humidity-Proofed"

To give you safe, sure protection against radio's deadliest enemy — dampness. Zenith's special wax-impregnating process effectively guards against the high humidity on shipboard that causes radios to lose sensitivity and thus fail to perform when you want or need them most.

Only $99.95* Less Batteries

Detachable Wavemagnet®

Swings up above the set for maximum efficiency. May be easily removed and attached to a window or porthole for better reception in boats, steel-shielded buildings, trains or planes. Special "pop-up" Waverod brings in Short-Wave stations with greater volume.

Spread Band Short-Wave Dial

Each of the 5 International Short-Wave Bands is "spread" on the large tuning scale. Each station is separated to make accurate pin-point tuning possible.

Push Button Band Selector

You need only push a button to tune in any band — Short-Wave or Standard Broadcast. The best time for listening to each of the 5 Short-Wave bands is listed right on the panel.

Your every hour afloat takes on new fun and thrills when you say "welcome aboard" to the "Trans-Oceanic." Now more powerful, more sensitive than ever, this newest version of the aristocrat of all portables brings you Standard Broadcast coast-to-coast PLUS international Short-Wave on 5 individual bands. Plays where ordinary portables fail — in boats, trains, planes, remote areas 'round the world.

"Humidity-Proofed" against loss of sensitivity due to dampness, this new Zenith® "Trans-Oceanic" is now lighter, easier to carry, and far lower in price than ever before! Works on Long Life Battery Pack and on AC or DC. See it at your nearest Zenith dealer's.

*West Coast and far South price slightly higher. Price subject to change without notice.

© 1950

Zenith Radio Corporation, Chicago 39, Illinois • Also Makers of America's Finest Hearing Aids

Universal Model G503.

G503
Universal
Companion to the G500 Trans-Oceanic

Introduction

A new member of the Universal line of AM portable radios was introduced in March 1950, about five months after the introduction of its companion, the G500 Trans-Oceanic. The Universal G503 replaced the egregiously ugly but entertaining "Batwing Door" or "Pop Open" Universal Model 6G801.

Versions

There were two versions of the G503 Universal: Black and Oxblood Brown. Both were covered in the now standard vinyl material, with the color being the only difference. The color of the plastic parts of each version matched the color of the cabinet covering.

Chassis

Black version - 5G40.
Oxblood version - 5G41.

Last Offered

The last advertisement for the G503 was in conjunction with an early H500 Trans-Oceanic advertisement in May 1951.

Design

The unique cabinet design of the G503 Universal is yet another attempt to solve one of the most difficult ergonomic design problems of portable radios in the tube era: how to remind the user to turn *off* the radio! This was a serious problem since vacuum tubes drew so much current and depleted the expensive batteries of the day rather quickly. Designers were doubly hampered since no small indicator lamps existed at that time which drew minimal current; having a dial light constantly on to remind the user to turn the radio off was self-defeating. Various strategies of pop-up red flags and disappearing dials were used by industrial designers to try to solve this problem.

Budlong and his staff developed the "flip-dial" concept to solve this problem; when the cabinet was closed (by rotating the dial forward and downward) the radio automatically shut itself off. This design concept is unique to the G503 Universal and to its slightly rounder sibling, the plastic-cased Zenith Tip Top Holiday portable.

The electronic design of the G503 is not particularly remarkable. The Zenith commitment to quality was embodied here in a fairly conventional "3-way" portable based on five miniature tubes. Advertising copy of the day claimed that the "giant 'Tip-Top' Dial with built in Wavemagnet swings up *above* the set for tip-top tuning ease, doubles the sensitivity...plus Wavemagnet concealed in the giant 'Tip-Top' Dial that swings high up, away from signal killing metal parts." There probably were some technical and operational advantages to raising the loop antenna substantially above the metal chassis; *doubling* sensitivity, however, was most likely advertising hyperbole.

Price

$49.95.

Companion

The G503 is a particularly beautiful companion for the G500 Trans-Oceanic.

Selected Print Advertising

NAT GEO	Apr 1950, F.A., Intro Ad. Great
HOLIDAY	May 1950, p.121, Intro Ad. Great
COLLIERS	July 1, 1950, p.11, Excellent
SAT EVE	July 15, 1950, p.55, same
SAT EVE	Nov 25, 1950, p.85, Excellent
SAT EVE	May 26, 1951, p.108, Color!

For a new thrill in radio— **at home or wherever you roam . . .**

New Zenith "Universal"
PORTABLE RADIO

*with exclusive "Tip-Top" Dial
and Wavemagnet*

HERE — from the world's leader in portable radios — is the most powerful standard broadcast portable in Zenith* history!

So sensitive is this stunning new Zenith "UNIVERSAL"* that it literally "reaches out" for distance. Brings in your favorite programs with amazing volume and tone beauty, even in places where many ordinary portables won't play!

The secret of this terrific performance is the stepped-up sensitivity made possible by Zenith's 3-Gang Tuning. Plus a new, more sensitive

Zenith-built Alnico 5 Speaker that gives tone richness and clarity comparable to many console speakers!

The "UNIVERSAL" plays instantly when you open the lid, turns off when you close it. Lift the lid, and there's the giant "Tip-Top" Dial, actually *above* the set for easier tuning. Wavemagnet* built right into lid also swings up, away from signal-killing metal parts.

Plays on its own long-life battery, and AC or DC. Handsome, sturdy, luggage-style case in buffalo-grained black or brown. See this tremendous new value now, at your Zenith dealer's! Only $49⁹⁵†

Less Batteries

New Zenith "TRANS-OCEANIC"

New edition of the world-famous standard and short-wave portable with overall performance superior to that of any other portable. More powerful, more sensitive than ever — yet lighter, easier to carry, and lower in price. Brings in Standard Broadcast, plus International Short Wave on 5 separate bands. "Tropic-Treated" against humidity. Plays on thrifty long-life battery, and AC or DC. $99⁹⁵†

Less Batteries

© 1950   *REG. U.S. PAT. OFF.

† West Coast and far South prices slightly higher. Prices subject to change without notice.
Over 30 Years of "Know-How" in Radionics* exclusively • Zenith Radio Corp., Chicago 39, Illinois
Also makers of America's Finest Hearing Aids

Trans-Oceanic Model H500.

Trans-Oceanic Model H500.

Trans-Oceanic Models G500 and H500.

Trans-Oceanic Model H500.

H500
The *Super* Trans-Oceanic

Introduction

The H500 was introduced with a press release on May 31, 1951, too late for the 1950-51 Christmas season and almost too late for the summer season so important for the sale of portable radios. There was no evidence in the advertising of what must have been a monumental struggle to give birth to this now (finally) "all-new" Super Trans-Oceanic. The summer of 1951, and the following Christmas season, witnessed a more extensive print advertising blitz than that afforded any Trans-Oceanic model before or since. Although production figures are unavailable, current evidence indicates that the H500 had higher annual production rates than any of the other Trans-Oceanic tube models.

Versions

There was only one version and one chassis, 5H40, for the Model H500. Note, however, that the only true military version of the Trans-Oceanic, the R-520/URR in Department of Defense nomenclature, was a modified version of the H500 (the R-520/URR is treated as a separate model in the next section).

Last Offered

In the 1953 Line.

Cabinet Design

The major appearance changes in the H500 represent a rather complete redesign of the basic suitcase portable. All of the changes, however, were entirely cosmetic styling. Industrial designer Budlong, with Commander McDonald's support, sought a more modern, softer 1950s look. The all black and gold ("Royalty of Radio") look was lightened with the use of gray paper behind a new gray Wavemagnet. Rounded corners on the front cabinet door, the Wavemagnet, and the main frame of the plastic front face also lent a softer look to the new Trans-Oceanic styling concept.

Electronic Design

Since the design of the "all-new" H500 entailed a new front face, push buttons and dial markings, as well as the changes already noted, the Zenith engineers were free to make the first changes to the spectrum coverage of the Trans-Oceanic since the pre-war introduction of the 7G605 Clipper. They abandoned the electrically bandspread 49 meter International Broadcast Band (5.950 MHz to 6.250 MHz) and added continuous shortwave coverage from 2 to 8 MHz, spread across two bands. This entailed the only major redesign of the RF Head undertaken during the tube years. The old "bandspread only" coverage completely ignored the regionally-oriented Tropical Broadcast Bands found in the 3 MHz and 5 MHz range, as well as many other forms of communication then found in the lower half of the shortwave spectrum.

The new "general coverage" bands were useful to many consumers. There is little doubt, however, that this rather major redesign was undertaken to satisfy the needs of Commander McDonald and his many yachting friends; weather radio and navigational stations of great importance to yachtsmen and professional sailors were then found in the now tuneable 2, 3 and 7 MHz regions of shortwave. The H500 was an instant success with the yachtsmen of the day; their testimonials and photos filled much of the print advertising for the H500.

Price

The introductory price of the H500 in the summer of 1951 was $99.95. By the Christmas season of 1951, the price had jumped up to the old price of the 8G005Y, $124.25. This price held throughout the H500 years.

Companion

Several of the early H500 advertisements show the G503 Flip Dial Universal continuing as the companion. A new Universal was introduced in the 1952 line (December 1951), the Model H503Y. It continued the Universal model name, was covered in Black Stag, and had a Wavemagnet in a hinged door; however, it was very curvilinear and could *not* be called a Companion to the H500 Trans-Oceanic. It was priced at $59.95. In June 1952, Zenith introduced yet another Universal, Model J504. This Universal had a case fabricated entirely from plastic but partly covered in Black Stag; there was also a brown version. Neither version of the Universal J504 could be considered Companions to the H500. In May 1953, Zenith introduced a final Universal, the L505, originally priced at $54.95. Although the L505 still had a removable Wavemagnet, it was not really related to the Trans-Oceanic of that day. The L505 survived until at least the 1956 line (then called the T505; this was the last use of the model name "Universal"). *Note:* Zenith also introduced a "wannabe" Trans-Oceanic, the (rare) L507 "Meridian," in the spring of 1953. The Meridian contained MW + 2 SW bands and retailed for $89.95; the H500 sold for $124.25. This receiver is not considered a true companion receiver to the Trans-Oceanic.

Selected Print Advertising

NAT GEO	June 1951, F.A., "World's Premiere," also other magazines
NAT GEO	Aug 1951, F.A., "No Other Radio Like This in All the World!"
HOLIDAY	Dec 1952, p. 138, "Unmatched Achievement of Zenith Quality"
NAT GEO	Feb 1953, F.A., "The Radio That Makes the World Your Oyster"
HOLIDAY	Jan 1954, p. 1, "Of course it has imitators,"Same ad, Dec 53

The last H500 advertisement was a full page color advertisement on page 1 of the January 1954 *Holiday*. This very expensive advertisement was the first full page ad for the H500 since July 1952, 17 months earlier. It appears to be an obvious manipulation of the market, designed to clear the Zenith warehouses of H500's at the full price of $124.25. When this ad was published, the all-new 600 Series Trans-Oceanic had been in production in Chicago for several months and would be introduced in less than a month.

Zenith quality in radio has been a watchword for thirty-two years. Zenith tone, selectivity, performance have challenged the radio industry again and again with important "firsts." Today, as in the past, Zenith radios, console and table models, clock radios, portables and the sensational Trans-oceanic, continue to spell *quality* first . . . to *last!*

This view shows the Trans-oceanic Portable assembly line. These girls, putting together intricate parts in perfect proportion, help to bring London to your fingertips as easily as Chicago.

Along the Trans-oceanic assembly line. This girl inspects Zenith's sensational portable before final assembly. Inspections such as these, at many points, maintain highest quality.

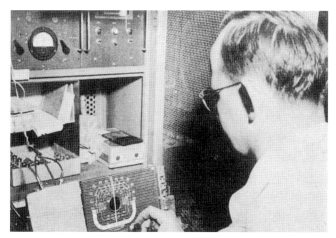

Balancing the Trans-oceanic. Audio testing to line up each individual part and put all of them into perfect balance for tone and selectivity before final testing.

The clock-radio assembly line. The "Q's" tell the quality story. Three weekly "Q's" earned in a row for quality production standards mean "coffee on the house" for each employee on the line.

Final assembly, testing and packing of Zenith clock radios. Over 850 clock radios roll off the line each day. These are the quality AM-FM models that "wake you up to music."

This is the <u>one</u>

ZENITH SUPER TRANS-OCEANIC
THE ORIGINAL
SHORT WAVE PORTABLE RADIO

- The only short wave portable with a twelve year record of performance around the world.
- The proved world portable with humidity-proofed chassis that works anywhere.
- Patented detachable Wavemagnet® antenna for maximum efficiency.
- The unequalled world portable that's your passport to 73 countries.
- The famous world portable that lets you hear ship-to-shore and ship-to-ship conversations, marine and weather reports, gets standard AM broadcasts, too!
- Zenith's finest portable with 7 tuning bands. Operates on AC, DC or Battery. As necessary in your home as a flashlight in case of electric power failure caused by air raid or other emergency.

In short wave radio this is the one, the original. It has never been equalled. And it actually costs less.

ZENITH
The royalty of television and RADIO®

Backed by 34 Years of "Know-How" in Radionics® exclusively
Also makers of fine hearing aids
Zenith Radio Corporation, Chicago 39, Illinois

COPR. 1953

ZENITH RADIO CORPORATION, Box NG63
6001 Dickens Avenue, Chicago 39, Ill.

Please send me your illustrated literature on Zenith Super Trans-Oceanic.

Name

Address

City _____ Zone _____ State _____

Pick up the world and take it with you

ZENITH SUPER TRANS-OCEANIC RADIO

Here's a radio to stir the blood of the man who's weary of humdrum days. It's the Zenith Super Trans-Oceanic Portable.

You can listen direct to foreign-wave broadcasts from all over the world. You can get ship-to-shore conversations, marine reports, aircraft communications, and the standard stations, too.

For the "man who has everything" or language students or a loved one in the service, there is no finer gift than the Zenith Trans-Oceanic Radio. Hear it at your Zenith dealer's.

There is only one genuine Trans-Oceanic Portable—originated by Zenith and never equalled. For your protection, insist on the set with the Zenith Crest inside.

7 Tuning Bands • Works on trains, planes, ships at sea and in steel buildings • **Humidity-proofed Chassis.** Black Stag Case. AC, DC, or Battery. Invaluable in case of electric power failure caused by an air raid or any other emergency.

ZENITH
The royalty of television and RADIO®

ZENITH RADIO CORPORATION, Chicago 39, Illinois
Backed by 34 years of "Know-How" in Radionics® Exclusively
Also makers of fine hearing aids

COPR. 1953

Ahora... Aun Mayor ALCANCE MUNDIAL!

Recepción mundial en 7 bandas cubriendo onda larga, 4 bandas internacionales de onda corta y 2 bandas contínuas de onda corta de 38 a 75 y 75 a 150 metros.

Modelo H500

NUEVO ZENITH
SUPER TRANS-OCEANIC
Para Ondas Largas y Cortas

Alcance Completo de Onda Corta en Bandas Ensanchadas de 16, 19, 25 y 31 metros . . . también en 38, 41, 49, 52, 60, 75, 90 y 120 metros.

Completamente a prueba de Humedad para protección contra pérdidas de sensibilidad y el deterioro de piezas vitales en cualquier clima.

Opera en Cualquier Parte—en trenes, aviones, a bordo de vapores, en edificios de acero, con cualquier fuerza "standard" CA, CD o con sus propias baterías.

El único radio de su clase en el mundo—hecho aun más versátil que nunca! Para alcance a través de océanos y continentes . . . para información directa desde las capitales bélicas y puntos estratégicos del mundo. Para traer a usted información y entretenimiento que radios ordinarios no pueden captar—no importa donde vaya ni lo que haga. Vea, escuche hoy mismo este nuevo sensacional Zenith Super Trans-Oceanic donde su distribuidor local!

GRATIS! ENVIE ESTE CUPON PARA RECIBIR DETALLES RESPECTO AL SUPER TRANS-OCEANIC!

ZENITH RADIO CORPORATION
6001 Dickens Ave., Chicago 39, Ill., E. U. A.

Nombre. . . . .

Dirección. . . . .

Ciudad. . . . . Estado. . . . .

País. . . . .

Trans-Oceanic Model R-520/URR.

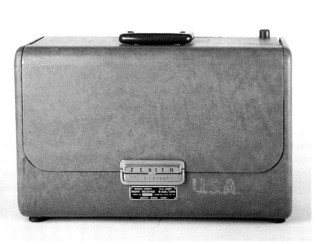

Trans-Oceanic Model R-520/URR.

Detail of military identification plate for the Trans-Oceanic Model R-520/URR.

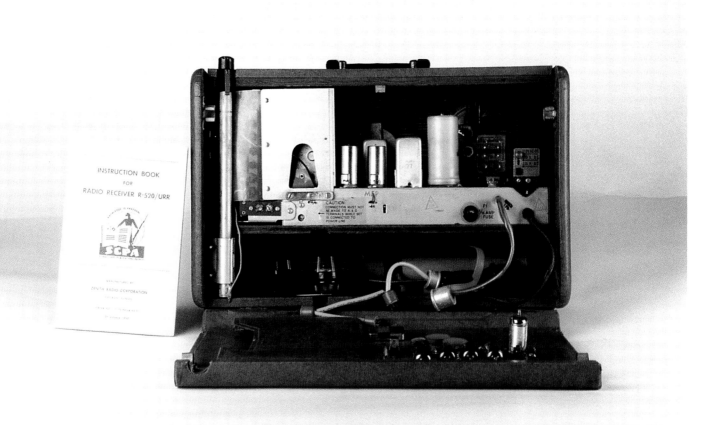

RADIO RECEIVER
R-520/URR
The Military Version

Introduction

The military version of the Trans-Oceanic was produced under contract No. 15175-P-52 with the Department of the Army; the units were fabricated during 1953 or very early 1954. These dates are supported by the publication date of Army Technical Manual (TM 11-877), published in January 1954. A copy of this Technical Manual was packed with each R-520/URR. With a contract date in 1952 and production in 1953, it is very probable that military procurement was motivated by the popularity of the 8G005Y and G500 models purchased by the troops for use in the field and their families at home during the Korean Conflict. The Korean Truce Agreement was signed July 27, 1953.

Number Produced

Although production figures were not located, personnel associated with Zenith manufacturing at that time remember only one production run and estimate that only about 5000 R-520/URR models were built.[2] Today, the R-520/URR is very rare. (*see also the* Endpiece)

Service

During the preparations for the Bay of Pigs Invasion of 1961, the U.S. military contracted with Zenith to produce either 250 or 500 containers for R-520/URR Trans-Oceanics. These containers were to be designed specifically to allow the radio to survive an "air drop."[3] [*Note*: There has been a rumor around the radio hobbies for a number of years that President Fidel Castro's favorite personal radio was a Zenith Trans-Oceanic--could there be a connection?]

It is probable that some R-520/URR models saw service in the early years of the Vietnam war; however, no accessible records exist which would confirm this supposition.

Versions

There was only one "military version" of the Trans-Oceanic built to strict military specifications: the highly modified H500 discussed here. One immediate indication of a "mil-spec" tube-type Trans-Oceanic chassis is that it contains no paper capacitors. (*see also the* Endpiece)

Please note that the USO organization (and possibly the services themselves) purchased a number of the 600 Series Trans-Oceanics. These radios usually had "Property of the U.S. Army" stenciled across the back of the cabinet. They were available for rental to service personnel and their families, along with other recreation equipment, at major military bases. These radios did not carry a military receiver designation. Other than the painted stencil, these Trans-Oceanics were identical to other leather 600 Series radios.

Cabinet Design

The R-520/URR was a highly modified Zenith Trans-Oceanic H500. From the exterior, four changes from the stock H500 are apparent. First, rather than the standard Black Stag, the cabinet is covered with brown "oilcloth." Second, all black and gray plastic H500 parts were replaced with dark and light brown respectively. Third, there was a standard aluminum military identification plate bradded to the cabinet just below the main front cabinet latch. Finally, there is a dial light (3 milliamp glow-type lamp) placed subtly behind the upper right corner of the rectangular dial escutcheon.

An inspection of the interior reveals a number of other modifications to the stock H500. All R-520/URR radios came with the Z-1 plug-in power supply module, allowing operation of the receiver with a number of different electrical systems at a variety of voltages. There was also a clip, mounted to the underside of the cabinet top, which suspended four different power plug adapters directly above the center of the chassis; an additional clip beside, and parallel to, the Waverod held an alignment tool. The very fragile RF head was also totally encased in a detachable sheet metal shield.

The interior surface of the back cabinet door contained a number of other accessories. There were clips to hold a spare dial light and spare fuses, a long metal clip to hold an entire spare set of vacuum tubes and a built-in deep pocket to carry the highly detailed 128 page military service manual, TM 11-877. All in all, the accessories provided with the R-520/URR, coupled with the extensive service manual, equipped the basic H500 for long and hard use in the field.

Electrical Design

The major electrical changes from the stock H500 were the provision of the dial light and a plug-in current regulator resistor. The dial light is a 105-120v .003 amp glow-type which indicates when the radio is on.

One of the serious weaknesses of the still relatively new miniature tube-based chassis introduced in the G500 was an oversensitivity to fluctuations of the AC power voltage. [See comments concerning this problem in the following 600 Series discussion.] The Zenith engineers selected a thermal resistor current regulator to eliminate this problem. Since it was glass-encased in a standard 9-pin miniature tube housing, it is almost indistinguishable from a vacuum tube.

The military specifications under which the R-520/URR was built were quite stringent, calling for the very highest grade of components available with no real concern for parts cost or ease of assembly. These specs also called for the elaborate extra protection of the various components. These "extras" caused the production line to run much more slowly than normal.[4] One wonders if this extra ruggedization was warranted given the exemplary service record of the stock 7G605 in World War II and the G500 in Korea.

Final Note

The Trans-Oceanic chassis was from the beginning protected from excessive humidity by a heavily painted chassis and wax impregnated coil inductors. Zenith and the Department of the Army went one step further with the R-520/URR, dipping the chassis in a special fungicide and humidity protective bath. This process left a tough yellow residue on all parts of the chassis, the electrical wiring and components. This same bath was used on other receivers procured for military use during those years, including the military version of the massive Hammarlund SP-600.

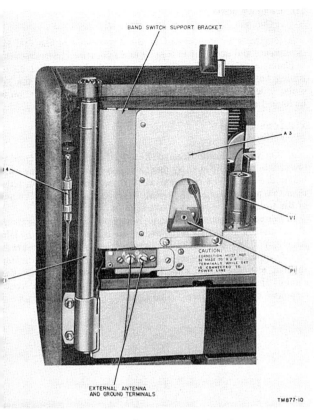

Antenna and ground terminals.

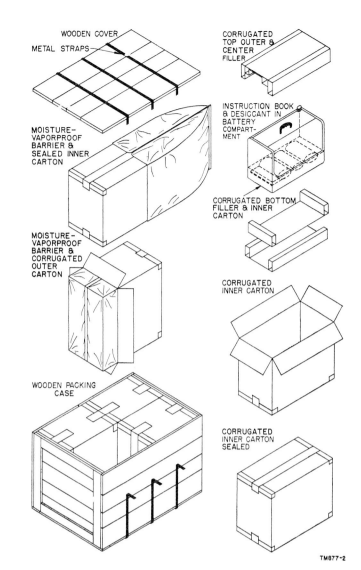

Packing and packaging of Radio Receiver R-520/URR.

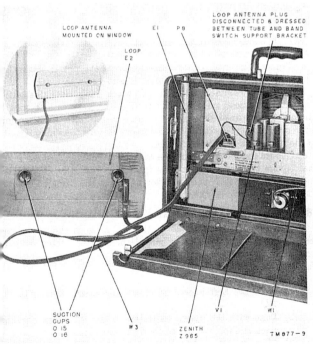

Extension cable connected to loop antenna.

Figure 1. Radio Receiver R-520/URR.

600 Series Trans-Oceanic

600 Series Trans-Oceanic

Leather version and Black Stag version 600 Series Trans-Oceanics.

600 Series Trans-Oceanic

THE 600 SERIES
Introducing the Slide Rule Dial

Introduction

The "600 Series" was introduced with the L600 model in spring 1954. Although there was an early "teaser" magazine advertisement in February 1954, the real advertising blitz was saved for the summer season with advertisements starting in national magazines in May 1954. The L600 marked the introduction of the "slide rule" dial on the Trans-Oceanic. This totally horizontal dial offered a very real improvement in tuning ease. The L600 could also be purchased "covered in genuine cowhide" for an additional $20. There were many other refinements in circuitry and in "user-friendly" details. *Unlike* the advertisements for the new design features introduced in the G500 and H500, the advertisements for the introductory 600 models were very explicit and discussed the new features in great detail.

Versions

There were six models produced in what the authors characterize as the "600 Series." In order of introduction, these were the L600, R600, T600, Y600, A600, and B600. With no styling changes, and very few changes to the electronics, these "models" were really all versions of the original L600. Due to the large amount of confusion and misinformation existing about various aspects of these six "models," a detailed discussion of each is presented at the end of this section.

Cabinet Design

This was the last Trans-Oceanic designed by Robert Davol Budlong and his associates, and it was a tour de force. The new design retained elements of the H500 cabinet--even the entire plastic front panel of the radio--although few people notice this even in side-by-side comparisons. The most noticeable change introduced with the L600 was the slide rule dial. This straightened horizontal arrangement allowed a longer and more easily readable dial for each band, thus simplifying tuning. When coupled with the electrical bandspread concept, this dial made the 600 Series Trans-Oceanic one of the easiest consumer radios to tune (until the advent of digital tuning 25 years later). There were a number of other design innovations developed by the Zenith/Budlong team to improve the usefulness of this last-of-the-tube Trans-Oceanics. These include:

Dial Light: A spring-loaded switch controlled a separate 1.5v battery
Power Cord Take-Up Reel
Attached Log Book: Reintroduced an attached log book last seen in the 1949 8G005Y
On/Off Indicator tag on volume control knob: An attempt to solve the battery drain problem caused by leaving the radio on in transport.
Small Wavemagnet: One improvement introduced in the 600 Series that is sometimes overlooked was the new smaller Wavemagnet. It was now a ferrite bar encapsulated in black plastic and removable from the top of the receiver cabinet.
Leather Case Option: The genuine cowhide-covered case was coupled with a change from the standard black and gray plastic cabinet parts to rich dark and light brown plastic to match the rich brown of the cowhide. The combination produced what many feel to be the most elegant Trans-oceanic ever offered to the public.

Electronic Design

The major change in the electronic design of the 600 Series chassis was the introduction of the 50A1 current regulator tube. It was necessary to add this tube since the 1.5-volt oscillator tubes were being driven to their design limits to perform at shortwave frequencies, and variations in AC line voltage sometimes caused the oscillator tube to "drop out" and the set to cease functioning. Variations in normal line voltage could also cause the oscillators to fluctuate a bit in frequency, affecting the accuracy of the dial or making stations sometimes seem to "drift" in frequency. This was not a constant problem and did not happen at all when the sets were operated from batteries. The inclusion of the 50A1 tube in the 600 Series models solved this problem.

Price

Price was $139.95 for Black Stag and $159.95 for Cowhide (first offering of leather). Prices held at $139.95/$159.95 (leather) throughout the offering of the 600 Series.

Commentary

We should note that the Hallicrafters Corporation, also of Chicago, had attempted to emulate the Trans-Oceanic by introducing the TW 1000 in 1953. The TW 1000 had a slide rule dial and a removable ferrite MW antenna equipped with suction cups, like the Wavemagnet. Further, the tube complement was identical to the G500 and H500. The styling was similar, but, from a 1990s viewpoint, not nearly as sumptuous as the Zenith radios. The slide rule dial RCA Stratosphere was very similar in design and appearance to the leather version of the L600 and was introduced three or four months before the L600. Long-time Zenith employees feel that the creation of the RCA Stratosphere was most likely an outgrowth of the lifelong enmity between the presidents of these two corporate giants.

Selected Print Advertising - 600 Series

HOLIDAY	Feb 1954, p.81, Introductory Advertisement	
NAT GEO	May 1954, F.A., Introductory Advertisement	
NAT GEO	Oct 1954, B.A., "Captain USN on camel" 1st offer of leather version	
HOLIDAY	Aug 1955, B.C., "Bhutan Tribesmen"	
NAT GEO	Dec 1955, F.A., Last Advertisement with Adm. McMillan	
NAT GEO	April 1956, F.A., "Yacht Race"	
NAT GEO	Sept 1956, F.A., "Elderly Beach Scene"	
NAT GEO	April 1957, F.A., Last major 600 Series Advertisement	

Commentary

Much misinformation exists among collectors as to exactly when each of the various 600 Series models was produced. There is also much misinformation as to what the various model number designations mean. The best remaining sources at Zenith are the records of the Service Department and the semi-annual "Spec Books" held in the corporate archives. The full details of Trans-Oceanic models and pertinent dates are presented as Figure 6-1 in the "Lineage" chapter.

NOTES ON 600 SERIES MODELS

Deciphering The Model Numbers

In the late 1940s, Zenith adopted a corporation-wide code that (internally at least) consisted of a trailing letter behind a model number indicating color. The code for the color black was "Y." Thus, a black L600 was more properly an L600Y. The corporate code for leather covering was "L"; thus, a leather-covered T600 was a T600L. Since the black model was *always* the primary Trans-Oceanic model, the trailing Y was often used only on internal corporate documents.

In the 600 Series, the preceding letters were assigned more or less on a yearly basis, in the order: L600, R600, T600, Y600, A600, B600.

Deciphering The Chassis Numbers

All tube models after the 8G005 had similarly patterned chassis numbers: 6x40y. The leading "6" indicated the number of tubes. This number was a 5 on the G500 and H500 chassis (5G40 and 5H40 respectively). The first letter of the chassis number usually indicated the model (Example: the G500 model chassis was the 6G40). The "40" (or in some cases, the "41") was an arbitrary number which identified the chassis. With the tube-type Trans-Oceanics, the 40 chassis indicated the standard Trans-Oceanic chassis. The 41 chassis was introduced with the 600 Series. The only difference between the two was the background color of the large slide rule: dial color of the 40 was black; the 41 had a brown dial and was used only in leather covered versions. A trailing letter in the chassis number gave various other specific information. The only trailing letter used in the tube-era Trans-Oceanics was a "Z" and indicated a change in the design of the power supply. This occurred in the Y model (chassis 6T40Z and 6T41Z) and the earlier models 8G005TZ1 and 8G005TZ2.

L600 (1954-55)

Introduced
May 1954 advertising blitz, with one early advertisement in February 1954 *Holiday*, page 81.

Last Offered
Last print advertisement was in the November 1955 *Holiday*, page 141.

Chassis
6L40 and 6L41.

Notes
The February 1954 *Holiday* advertisement is full page "announcing" the Super *Deluxe* Trans-Oceanic and details the many new features. Beneath the main photo of the set is "Designed by Robert Davol Budlong."

Although the L600 had no front phone jack, by December of that first year, some advertisement photos show a front phone jack.

R600 (1955)

Introduced
Corporate records indicate sales in the 1955 Line, while Service Department publications indicate production sometime in 1954. However, advertising for the L600 continued throughout 1955; R600 advertisements ran from May, 1955 to April, 1956 and overlap *both* the L600 and the T600 models.

Last Offered
The last advertisement was April 1956, followed in May 1956 by the first advertisement for the "Y600."

Chassis
6R40 and 6R41.

T600 (1955)

Introduced
1955 Line.

Last Offered
1956 Line.

Chassis
6T40 and 6T41.

Notes
The Trans-Oceanic advertisement in *LOOK*, November 1, 1955, page 9, calls the radio a "T600" and mentions a "phono-jack." The only print advertisement for the T600 was in *Collier's*, December 9, 1955, page 15.

The Y600, A600 and B600 models came with handle brochures calling them "T600." The Y600 used the 6T40 and 41 chassis with modifications to the power supply.

Price
December 1955, prices still $139.95/$159.95 (cowhide).

Y600 (1956-57)

Introduced
Advertisements show the Y600 was introduced for summer 1956. *Collier's*, May 11, 1956, page 13, calls it a "Y600."

Last Offered
At end of 1957 Line.

Chassis
6T40Z and 6T41Z.

Price
$139.95/$159.95; the prices were still the same in November, 1956.

A600 (1958)

Introduced
There were no national print advertisements for the A600 model, since all attention was focused on the new all-transistor Royal 1000. A600 was the model designation for the 1958 model.

Last Offered
1958 Line.

Chassis
6A40 and 6A41.

B600 (1959-62)

Introduced
In the 1959 Line.

Last Offered
Closed-out during the 1962 model year.

Chassis
6A40 and 6A41.

Notes
The Portable Radio in American Life, p. 197, notes that the B600 was the last tube model portable radio made in the United States. Magazine advertising only shows the Royal 1000/1000-D Trans-Oceanic in the period 1957-1960. There was no print advertising for the A600 or B600.

they tuned in the world from
"TREASURE ISLAND"
with the world's most <u>treasured</u> radio!

ONLY ZENITH HAS <u>ALL</u> THESE QUALITY FEATURES

Earphone jack; phono jack; detachable Wavemagnet® antenna with rotating directional feature; 5-foot pop-up Waverod antenna for Short Wave; log chart compartment; Reelaway power cord; battery saver switch; Zenith quality speaker with Alnico-5 magnet.

On beautiful Isle de Pinos, believed to be Stevenson's *Treasure Island*, the Daun Yeagleys tune in foreign lands on their Zenith TRANS-OCEANIC® radio.

Peak Reception! To boost sensitivity aloft, the Yeagleys usually attached the famous Zenith Wavemagnet® antenna to window.

Great "Co-Pilot"! The TRANS-OCEANIC radio is the companion of flyers, diplomats, newsmen, explorers and scientists the world around!

FOR HOME DEFENSE: A Zenith battery-operated portable is as necessary in your home as a flashlight in case of power failure or air raid!

THE FAMOUS ZENITH
TRANS-OCEANIC® PORTABLE RADIO

proved on the sands of paradise
...powered to bring in the world!

ZENITH SHORT WAVE RADIO once again proved its incredible *power* by keeping this flying couple in touch with the world during their recent Caribbean trip!

Mr. and Mrs. Daun Yeagley of Minerva, Ohio, used their TRANS-OCEANIC radio for entertainment, weather reports and radio bearings from the U.S. and other countries. Reception was excellent both in their light plane and on remote islands, including the high point of their flight: historic Isle de Pinos, 35 miles S.E. of Cuba.

You, too, can keep up with the world with this famous radio—the world's *only* 15-year proved Short Wave portable. It covers long-distance Standard broadcasts, as well as international Short Wave bands; marine, weather and amateur Short Wave coverage bands; and ship-to-ship and ship-to-shore reception. It operates on planes, ships, trains, in steel buildings; and it is treated to resist humidity. AC/DC or battery operated. In rugged black Stag, $139.95*; luxurious top grain cowhide, $159.95*.

**THE QUALITY GOES IN
BEFORE THE NAME GOES ON**

The Royalty of RADIO, TELEVISION and PHONOGRAPHS
Backed by 37 years of leadership in radionics exclusively
ALSO MAKERS OF FINE HEARING AIDS
Zenith Radio Corporation, Chicago 39, Illinois

*Manufacturer's suggested retail price, not including batteries. Slightly higher in Far West and South. Prices and specifications subject to change without notice.

The schooner "Constellation" —Class A winner in the 1955 Trans-Pacific Yacht Race.

It's a sailor's radio!

If it's good for a sailor, it's good for everyone... there's no place _wetter_ than the ocean!

Zenith TRANS-OCEANIC® Short Wave Radio keeps "Constellation" in touch with world during historic ocean race!

ZENITH Short Wave radio wrote yachting history last July when it sped across the Pacific from Los Angeles to Honolulu aboard the fastest Class A schooner in the 1955 Trans-Pacific Yacht Race!

Frank Hooykaas, owner-skipper of the 75-foot "Constellation," used his Zenith TRANS-OCEANIC portable radio constantly to tune in network broadcasts from the U. S. and Hawaii, and to take precise radio bearings when approaching the island of Oahu.

You, too, can tune in the world with the Zenith TRANS-OCEANIC portable radio, or get the best long distance domestic reception on standard broadcast! It also covers international Short Wave bands; marine, weather and amateur Short Wave coverage bands; plus ship-to-ship and ship-to-shore reception. Students can tune in foreign lands and learn languages as they are _actually spoken!_

This 15-year-proved Short and Standard Wave radio operates on ships, trains, planes and in steel buildings. Tropically treated against high humidity to prevent loss of sensitivity. AC/DC or battery operated. In black stag, $139.95*; brown cowhide, $159.95*.

IMPORTANT! A Zenith battery-operated portable is as necessary in your home as a flashlight in case of power failure caused by air raid or other emergency!

Skipper Frank Hooykaas (second from left) and crew listen to vital weather information on their Zenith Trans-Oceanic Radio. Since radio is equipped with an earphone jack, it can be played without disturbing sleeping crew members.

THE QUALITY GOES IN
BEFORE THE NAME GOES ON

The Royalty of RADIO and Television®
Backed by 37 years of leadership in radionics exclusively
ALSO MAKERS OF FINE HEARING AIDS
Zenith Radio Corporation, Chicago 39, Illinois

*Manufacturer's suggested retail price not including batteries. Slightly higher in Far West and South._
Prices and specifications subject to change without notice.

93

Zenith Announces the Greatest News in Shortwave Portable Radio!

... Since the original Zenith TRANS-OCEANIC made shortwave history 13 years ago.

ALL NEW ZENITH SUPER DELUXE TRANS-OCEANIC PORTABLE

Tropically treated against humidity, it's ideal for Sportsmen, Outdoormen, Travelers ...

Now! 300% More Sensitive

New improved patented detachable Wavemagnet® Antenna pulls radio waves out of the air like a magnet, *anywhere*—on ships, on planes, on trains ... and only Zenith has it!

New Deluxe! Zenith Voltmatic Regulator—only Zenith has it ... automatically maintains constant power flow, extends tube life, lets radio play in remote places with haphazard power supply.

New Deluxe! Zenith Reelaway Power Cord—only Zenith has it ... pulls out from side of case, plugs in for AC or DC power, springs back when not in use.

New Deluxe! Zenith Log-Chart Compartment ... includes complete weather and marine information. Lists data on ALL major shortwave stations. Room for your own log too.

New Deluxe! Zenith On-Off Indicator ... provides extra safety check against power-loss when the set is not in use.

New Deluxe! Zenith's International Tuning Dial ... ... easier reading, simplifies locating and tuning stations all over the world.

New Deluxe! Zenith Spring Button Dialite ... illuminates entire face of dial, lets you tune stations in pitch darkness.

Only the Zenith TRANS-OCEANIC portable has the 13 year proved, tried and tested *reputation* in every corner of the world!

PLUS THESE ZENITH POWERIZED FEATURES

• Long Distance Chassis brings programs from dozens of distant countries. • Tropically treated against humidity to prevent loss of sensitivity. • Super sensitive electrical spread band tuning brings ship-to-ship conversations, Marine and Weather reports, amateur broadcasts, popular programs from all over the U. S. • Exclusive Radiorgan® Tone Control gives choice of 16 different tonal combinations. • Works on AC, DC or Long-life batteries.

Model above in durable Black Stag. Also available in luxurious top-grain cowhide.

ZENITH
The royalty of television and **RADIO**®

Backed by 36 years of Leadership in Radionics Exclusively
Zenith Radio Corporation, Chicago 39, Illinois

See your Zenith Wholesale Man Now!

Copr. 1954

Announcing the New Zenith
Super <u>Deluxe</u> TRANS-OCEANIC Portable

7 NEW features added to the World's only 13 year proved Shortwave Portable Radio

1. New Powerized Detachable Wavemagnet® Antenna ... plus powerful new circuits increase sensitivity on the standard broadcast band up to ... three times!

2. New International Tuning Dial permits far easier reading! Simplifies locating and tuning stations all over the world on shortwave or standard broadcast bands.

3. New Spring-Button Dialite illuminates entire face of dial. You can easily tune in your station in pitch darkness. Automatic release avoids excessive battery drain.

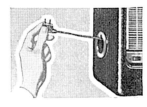

4. New Reelaway Power Cord reels out from side of case. Plugs in for AC or DC power. Springs back when not in use, readying set for battery operation.

5. New Voltmatic Regulator automatically maintains constant power flow through set, regardless of fluctuations at power source. Extends tube life!

6. New Log-Chart Compartment includes 24 pages full of complete weather and marine information. Lists data about all major short-wave stations in the world.

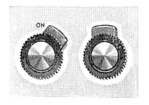

7. New On-Off Indicator further guards against excess battery drain ... provides extra safety check against power loss when set is not in use.

ZENITH SUPER DELUXE <u>TRANS-OCEANIC</u> RADIO

Model shown in durable Black Stag, $139.95. Also available in luxurious top-grain cowhide at additional cost.*

Zenith POWERIZED Features!

- Tropically treated against humidity, to prevent loss of sensitivity. AC, DC or long-life batteries.
- Long Distance Chassis brings you programs from dozens of different countries.
- Super-sensitive electrical spread-band tuning ... brings you ship-to-ship conversations, marine and weather reports, amateur broadcasts, popular programs from all over U. S.
- Exclusive Radiorgan® Tone Control gives you choice of 16 different tonal combinations.

One of these battery-operated portables is as necessary in your home as a flashlight in case of power failure caused by air raid or other emergency.

ASK ANY ZENITH OWNER!

The royalty of television and RADIO

Backed by 35 years of Leadership in Radionics Exclusively
ALSO MAKERS OF FINE HEARING AIDS
Zenith Radio Corporation, Chicago 39, Illinois

*Manufacturer's suggested retail price (subject to change) not including batteries. Slightly higher in Far West and South.   COPR. 1954

600 Series, Leather Version.

Famous correspondent keeps on top of fast-breaking world news with Zenith Short-Wave Radio

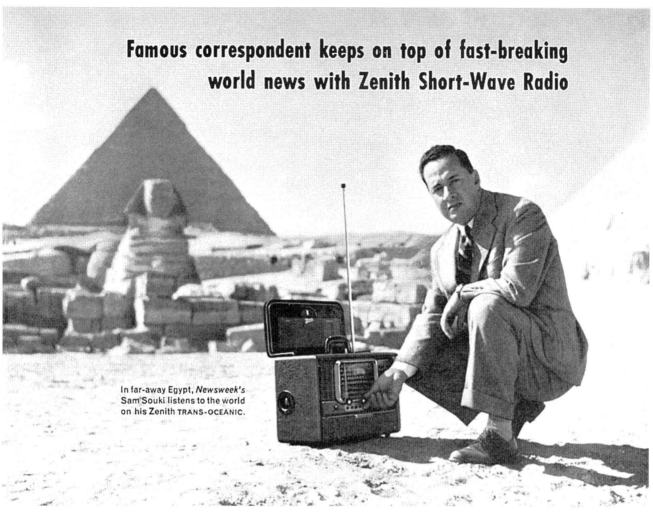

In far-away Egypt, *Newsweek's* Sam Souki listens to the world on his Zenith TRANS-OCEANIC.

THE INCOMPARABLE ZENITH TRANS-OCEANIC® PORTABLE RADIO

Proved in the Land of the Pharaohs
Powered to bring in the World

Mr. Sam Souki, *Newsweek's* famous Middle East correspondent, must keep on top of world news! That's why he uses the Zenith TRANS-OCEANIC Short Wave Portable Radio.

You too, can keep up with world news with this fabulous radio, or get the best long distance domestic reception on *Standard* Broadcast, as well. It also covers international Short Wave bands; marine, weather and amateur Short Wave coverage bands; and provides ship-to-ship and ship-to-shore reception!

This 15-year-proved TRANS-OCEANIC radio operates on ships, trains, planes and in steel buildings. It is built for rugged, outdoor use, and treated to resist high humidity.

The companion of newsmen, scientists, world explorers and sportsmen everywhere, the TRANS-OCEANIC'S list of owners reads like an international "Who's Who"! Operates on AC, DC or batteries. In black Stag, $139.95*; top-grain cowhide, $159.95*.

FOR FOREIGN LANGUAGE STUDENTS: This superb Short Wave radio tunes in foreign lands, lets you hear languages *as they are actually spoken!*

FOR HOME DEFENSE: A Zenith battery-operated portable is as necessary in your home as a flashlight in case of power failure or air raid.

 THE QUALITY GOES IN
BEFORE THE NAME GOES ON

 The Royalty of RADIO and Television®
Backed by 37 years of leadership in radionics exclusively
ALSO MAKERS OF FINE HEARING AIDS
Zenith Radio Corporation, Chicago 39, Illinois

Manufacturer's suggested retail price not including batteries. Slightly higher in Far West and South. Prices and specifications subject to change without notice.

Royal 1000 Series Trans-Oceanic.

Royal 1000 Series Trans-Oceanic.

Royal 1000 Series and 600 Series Trans-Oceanics.

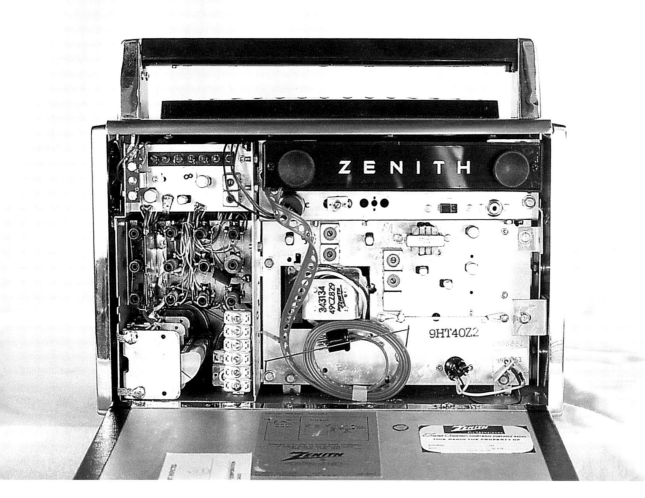

Royal 1000 Series Trans-Oceanic.

ROYAL 1000 SERIES
Introducing the
ALL TRANSISTOR
Trans-Oceanic

Introduction

The Royal 1000 was introduced in December 1957 in the last big advertising blitz implemented for any Trans-Oceanic. The headline of most of the early print advertisements was:

"Introducing...the World's Most Magnificent Radio...New *All Transistor* (TUBELESS) Trans-Oceanic!"

The Royal 1000 was to be the last "cutting edge of consumer electronics technology" radio for Commander McDonald and Zenith. It was perfected over an intense two-year period of engineering development; Zenith tradition has it that McDonald was determined to maintain the position of the Trans-Oceanic as the world's best portable radio. It was also determined to maintain its commitment to excellent reception while traveling (providing a removable Wavemagnet) and rugged construction for use in the field. For McDonald and his designers, these commitments meant an all-metal cabinet with doors which completely covered the front face of the radio and an unusually thick metal chassis.

Due to the untimely death of industrial designer Robert D. Budlong and the dissolution of his firm, the Chicago industrial design group of Mel Boldt and Associates was retained to design most Zenith consumer products after 1956. The Zenith engineers produced a number of prototypes of the "new all-transistor Trans-Oceanic" only to have each rejected as too bulky by Commander McDonald. He put the team back to work to produce the absolutely most compact Trans-Oceanic then possible. The result was the Royal 1000.

Versions

ROYAL 1000

Introduced
December 1957 for the 1958 Line. There were six different chassis for this model (refer to Figure 11-3). This large number of chassis reflects the continuing rapid development of new transistor technology.

Last Offered
The 1963 Line.

ROYAL 1000-D

Introduced
A July 3, 1958, press release introduced the Royal 1000-D as a version of the basic Royal 1000, but with a $25 option providing a 9th band (long wave) for "CAA weather/navigation." The Royal 1000 and the Royal 1000-D were advertised and sold simultaneously. There were at least seven Royal 1000-D chassis produced.

Last Offered
The 1962 Line.

Price
$275.

ROYAL 1000-1

Introduced
In the 1964 line, the "-1" denoted an input port for a 12v DC outboard power supply. The "-1" models of the Royal 2000 and Royal 3000 series were announced at the same time in a press release on December 2, 1963.

Last Offered
The Royal 1000-1 was last shown in the 1968 line, although it may have continued until the introduction of the Royal 7000, in April, 1969.

Chassis
Two chassis were produced: 9HT40Z2 and 9HT40Z8.

Cabinet Design

Even though the Royal 1000 was a completely new Trans-Oceanic in all respects, it owed much to the design ideas developed for the tube models by R. D. Budlong. First, the Mel Boldt and Associates designers of the Royal 1000 retained the concept of utilizing a hinged door to protect the delicate front panel of the receiver while in transit. In fact, this idea remained a "signature" of the Trans-Oceanic designs throughout their production. The commitment to electrically bandspread, separate shortwave bands focused on International Broadcasting, was also retained, as was the continuous coverage from 2 MHz to 8 MHz, primarily for yachtsmen and shortwave hobbyists. In fact, the spectrum coverage of the Royal 1000 was exactly the same as that of the H500 and 600 Series tube models.

The dial and bandswitching arrangement, however, was all new. The nine frequency bands of the dial were printed in slide rule fashion on a translucent plastic cylinder which was lit from its interior. This dial cylinder was mounted on the same shaft as the new rotary bandswitch. Thus, as the operator turned the bandswitch, the dial rotated automatically to expose the proper dial markings for the band in use.

The primary designer of the Royal 1000/1000-D cabinet was Anthony J. Cascarano, a young designer with Boldt and Associates.

Materials

The design team adopted a number of relatively new materials for the cabinet and front face. Long-term, some of these decisions turned out to have been unfortunate. Steel, with a thickly plated polished chrome finish, was chosen for much of the frame of the cabinet. A similar brushed chrome finish was used on the upper portions of the cabinet sides and top. Both finishes have deteriorated badly over time, with major pitting of both types of surfaces now being common. Further, and more disastrously, many receivers now exhibit brushed chrome finishes that are delaminating from the steel substrate, resulting in a blistered appearance (refer to the chapter on restoration for further discussion of this problem).

The selection of plastics was considerably more successful. The crystal clear acrylic molding of the dial escutcheon has maintained excellent color and clarity over the years. The first production runs of the Royal 1000 used black "Genuine Cowhide" covering (rare). All later versions of the Royal 1000 used black plastic to cover the steel cabinet: this dense, but flexible, sheet plastic material has held up well.

Early versions of the Royal 1000 also used a fixed internal battery box of clear rigid acrylic plastic. This box proved too fragile: later Royal 1000s were equipped with a removable battery box of flexible plastic.

Price

$250 throughout for the Royal 1000, $275 for Royal 1000-D. The cost of the Royal 1000 was reduced from $250 to $199.95 in April 1963, after the introduction of the $275 Royal 3000.

Companion

Royal 2000 AM/FM portable was intentionally designed as a companion for both the Royal 1000 Series and the Royal 3000 Series Trans-Oceanics.

Selected Print Advertising

HOLIDAY	Dec 1957, p.219, Intro Advertisement. Covers the main features. $250 price
NAT GEO	March 1958, F.A., Intro Advertisement "The World's Most Magnificent Radio!"
NAT GEO	July 1958, F.A., "Powered to Tune in the World!" (Also repeats in Sept., 1958)
HOLIDAY	Dec 1958, p.211, "World's Most Magnificent Gift!"
NAT GEO	May 1959, F.A., Father, son and globe in sumptuous family library
NAT GEO	June 1960, F.A., Air travel
NAT GEO	Dec 1960, F.A., Royal 2000-"Give the elegance and performance of Zenith"
HOLIDAY	Dec 1960, p.54, Husband and wife at Christmas tree ("She Gave You a Zenith!")
HOLIDAY	Oct 1961, p.154, "World's Most Powerful, Most Beautiful, Most Exciting"

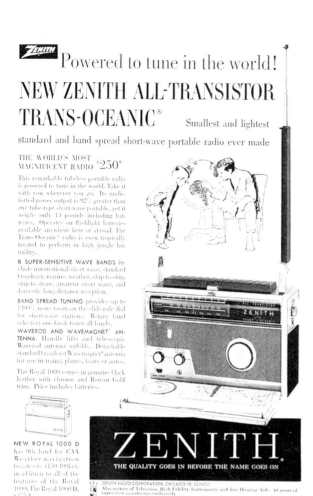

NEW! The world's most magnificent radio

ZENITH *All-Transistor* (tubeless)
TRANS~OCEANIC®
STANDARD AND BAND SPREAD
SHORT-WAVE PORTABLE RADIO

Performs where others fail...

yet it's the world's

lightest and smallest

Band Spread

Short-Wave

Portable Radio

*$250**

POWERED TO TUNE IN
THE WORLD...8 WAVE BANDS!

You'll carry this amazing new short-wave radio as you would a fine imported camera. It's engineered as soundly as a shockproof watch to perform where others fail. This superb radio is tubeless, needs no bulky "B" batteries or AC/DC power supply. Operates on flashlight batteries (available anywhere, here or abroad) for only a fraction of a cent per hour.

92% greater power output than tube type portables

Reception includes Standard Broadcast, International Short Wave, Marine, Weather, Ship-to-Ship, Ship-to-Shore, and Amateur Short Wave. Plays in planes, trains, autos, ships, steel buildings. Tropically treated to perform even in high jungle humidity.

This is the *All-Transistor* version of Zenith's original Trans-Oceanic® Radio ... the companion to royalty, explorers, yachtsmen, world-travelers. Its list of owners reads like an international "Who's Who."

REMEMBER...
a battery-powered portable is as necessary in your home as a flashlight in case of power failure caused by air raid or other emergency.

*Manufacturer's suggested retail price, including batteries. Price and specifications subject to change without notice.

QUALITY BY ZENITH
The Royalty of Radio

The quality goes in before the Zenith name goes on
ZENITH RADIO CORPORATION • CHICAGO 39, ILLINOIS
America's pioneer in fine radios for the home. Also makers of Television, High-Fidelity Instruments, and fine Hearing Aids.

Royal 2000 Series Trans-Symphony.

ROYAL 2000 SERIES
Not a Trans-Oceanic

Introduction

December 1960 for 1961 Line.

Versions

Royal 2000: 1961, 1962 and 1963 Lines.
Royal 2000-1: 1964 Line (and most likely 1965 and 1966 Lines). Refer to comments in the Portrait of the Royal 3000 concerning "-1" models.

Last Offered

The end date is not currently known. The Royal 2000-1 was probably closed out at the end of the 1966 Line.

Chassis

11ET40Z2 (used for both Royal 2000 and Royal 2000-1).

Cabinet Design

The "Trans-Symphony" Royal 2000 was designed by Gordon Guth, a young designer at Mel Boldt and Associates. Guth was asked to develop a design which would "fit in" to the Trans-Oceanic family. Since Zenith as a corporation was committed to high quality audio, the cabinet was to be large and sturdy to act essentially as a sound box for the very large, heavy 5" X 7" oval speaker. The easy-to-read dual dials moved on the same dial cord, in parallel with each other. Rather than resting in the center or lower portion of the cabinet, Guth flipped the chassis upside down and attached it to the underside of the top of the cabinet. This unusual location cleared space for the large speaker magnet and opened the cabinet interior for better acoustical fidelity. The Royal 2000 Trans-Symphony is a beautiful example of the early 1960s "clean look."[5]

An extra small dot may be seen just above and to the left of the "1" in the "100" on the FM dial of the Royal 2000. This extra dot is found on almost all Zenith FM dials from the 1950s and 1960s and marks the dial location of Zenith's own WEFM FM station in Chicago.[6] This dot is also found on the Royal 3000 Trans-Oceanic. Incidentally, the WEFM call letters were chosen to match the initials of Eugene F. McDonald, Jr.[7] By all rights, the Royal 2000 "Trans-Symphony" should have been called a "Universal." Unfortunately, that legendary model name died with the tube era.

Electronic Design

Inasmuch as the Royal 2000 was the first American AM-FM portable radio, the circuitry is innovative by definition. The design used eleven transistors, rather lavish when compared to the Royal 1000s nine. The tuning circuitry was unusual: the AM broadcast band was tuned by a conventional ganged tuning capacitor, but the FM band was tuned by three ganged permeably tuned oscillators (PTO's). These devices were normally only used on expensive communications equipment where rock-solid frequency stability was required.

Price

$189.95 at introduction in December 1960, dropping to $149.95 by October 1961.

Companion

Royal 1000 and Royal 3000 Series Trans-Oceanics.

Selected Print Advertising

NAT GEO	Dec 1960, F.A., Introductory Advertisement.
HOLIDAY	Dec 1961, p.154, "For People Who KNOW What They Want!"

New – from the world leader in FM !

Zenith proudly presents
America's first all-transistor
Portable FM/AM Radio

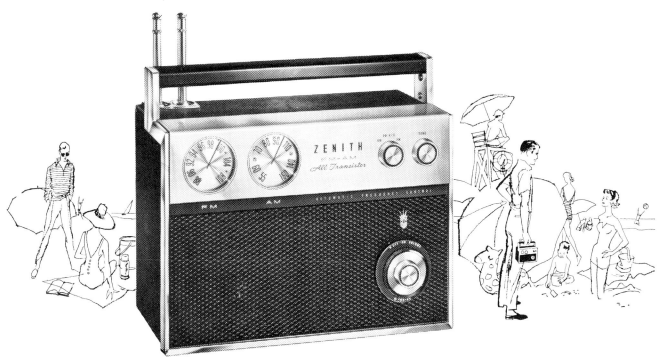

Engineered with watchmaker's precision, magnificently

styled, Zenith's new cordless Trans-Symphony Royal 2000

has richer, fuller tone – greater power and sensitivity –

than any portable radio of its kind ever made !

Now add the pleasure of FM to your *outdoor* listening. Zenith's new Trans-Symphony portable operates on ordinary flashlight batteries. Pours out rich brilliant tone from its 5″ x 7″ speaker. Like the finest table model FM/AM receivers, Zenith's new Trans-Symphony has Automatic Frequency Control for drift-free FM listening, broad-range tone control, precision Vernier tuning, Zenith's famous long-distance AM chassis. Three built-in antennas: a Wavemagnet® AM antenna, a concealed FM antenna, plus a telescoping FM dipole antenna. Weight: 11¾ pounds. Dimensions: 10 3/32″ high (including handle), 4⅞″ deep, 11⅝″ wide. Black Permawear covering, brushed aluminum and chrome plate trim. The Trans-Symphony Royal 2000, $189.95.*

Quality-built in America by highly skilled, well-paid American workmen.

ZENITH RADIO CORPORATION, CHICAGO 39, ILLINOIS. IN CANADA: ZENITH RADIO CORPORATION OF CANADA LTD., TORONTO, ONT. The Royalty of television, stereophonic high fidelity instruments, phonographs, radios and hearing aids. 41 years of leadership in radionics exclusively.
Manufacturer's suggested retail price, without batteries. Prices and specifications subject to change without notice.

ZENITH

The quality goes in before the name goes on

Royal 3000 Series Trans-Oceanic.

Royal 2000 Trans-Symphony and Royal 3000 Series Trans-Oceanic.

Royal 3000 Series Trans-Oceanic.

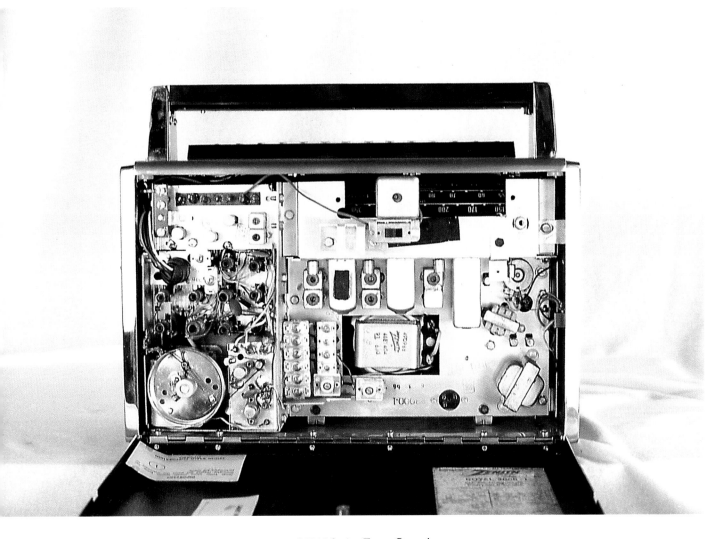

Royal 3000 Series Trans-Oceanic.

ROYAL 3000 SERIES
Introducing FM to Trans-Oceanics

Introduction

The Royal 3000 Trans-Oceanic was announced with a press release on November 16, 1962, for the 1963 Zenith line. With increases in FM transmitter power and the introduction of rock music to the FM band, FM became a feature that consumers expected on new portable radios. Zenith management and engineers sought to meet this new market by updating the FM-less Royal 1000/1000-D with a new FM sub-chassis. The only space available in the Royal 1000 cabinet was that occupied by the removable Wavemagnet. This last vestige of the original Trans-Oceanic Clipper was abandoned and a new FM sub-chassis was inserted. Room in the cabinet was so limited that it was necessary to add a curved back door which bulged enough to add almost an inch of front-to-back depth to the lower portions of the cabinet.

Special Note

The Royal 3000 was introduced as "The Ultimate in Personal Radios." The one millionth Trans-Oceanic was manufactured in early August 1964, in Chicago Plant #2. Plans were made to give this special set to then-President Lyndon B. Johnson; records in Zenith archives do not indicate whether this presentation ever took place.

It should be noted that the advertising campaign in national magazines for the Royal 3000 was minimal when compared to those of previous models. Examination of the print advertising of the entire Zenith line in the mid-1960s reveals several trends. First, there was a general de-emphasis on radio print advertising in that time period. This was both a reduction in absolute number and in the size of advertisements placed in popular magazines. It appears that Zenith's competitors may have also reduced their print advertising during this time period; however, these competitors appear to have done so by a smaller percentage. The balance of print advertising shifted to Zenith's American competitors and to the newly emergent multi-national corporations from Japan. The reasons for this shift are unknown. Further, the content of the smaller amount of print advertising that Zenith did purchase, even in *Holiday* and *National Geographic*, was decidedly non-radio-oriented, the focus being on console radio-phonographs and on television. A very few half page advertisements were purchased for the Royal 3000 in national magazines. The only full page ad known to have been purchased was in the radio hobby publication, *World Radio and TV Handbook*, 1965 edition.

The design decisions (a "band-aid" update of the Royal 1000-D) and the advertising decisions concerning the Royal 3000 seem, in retrospect, to have been poor judgement. It cannot be a total coincidence that Commander McDonald had died in 1958, and, for the first time, was not intimately involved in the design of "his radio".

Chassis

12K40Z3 (Royal 3000), 12KT40Z3 or 12KT40Z8 (Royal 3000-1).

Versions

Since all Royal 3000 models included the Long Wave Band, the nomenclature 1000/1000-D was dropped. There was, however, a Royal 3000-1 version (refer to the comments regarding "-1" models below).

Last Offered

The 1971 Line.

Cabinet Design

There were no major changes in the cabinet styling of the Royal 3000 other than the bulged back. The lower portion of the front panel was slightly rearranged, and a large metal and plastic logo was attached to the lower front portion of the outer side of the front cabinet door proclaiming the new FM capability. This face lift was done by designers at Mel Boldt and Associates, the firm that designed each of the cabinets of the solid-state Trans-Oceanics.

Electronic Design

The circuit design of the Royal 3000 approximates the Royal 1000-D, with an additional transistor to provide more IF amplification and an innovative two-transistor sub-chassis for FM RF amplification and mixing.

Price

The suggested list price of the Royal 3000 hovered around $275 until the Royal 7000 Series was introduced. The Royal 3000s price was then reduced significantly, and it was carried as the less expensive Trans-Oceanic model for several years.

Selected Print Advertising

HOLIDAY	March 1963, p.109, "Zenith Now Adds FM" small vertical ad
NAT GEO	May 1968, B.A., "... and FM, Too," half page horizontal ad
NAT GEO	Nov 1968, B.A., Same Dec., 1968, Mar. 1969. [Was this blitz to get rid of 3000s? The 7000 was coming in April, 1969.]
WRTH	1965, Full page

Special Note:
The Royal 1000-1,
Royal 2000-1
and Royal 3000-1

These models were introduced in the 1964 Line, the "-1" denoting an input port provided on the left side of the cabinet for a 12v DC power. These radios came with a Zenith plug-in "battery eliminator"-type external power supply. All three were introduced in a press release on December 2, 1963, for the 1964 Zenith Line.

Powered to tune in the world and FM, too ...Zenith's famous 9-band Trans-Oceanic radio

Zenith's world-famous Trans-Oceanic® portable is the finest solid-state radio built. Super-sensitive reception for AM, international shortwave, longwave and full-carrier AM amateur broadcasts, ship-to-shore and ship-to-ship, marine and FAA weather services. And, glorious FM, too! See the *Trans-Oceanic,* Model Royal 3000-1—at your Zenith dealer's. **ZENITH** ®

The quality goes in before the name goes on

Powered to tune in the world, and FM, too... Zenith's famous 9-band Trans-Oceanic radio

Here is the world's finest solid-state portable radio. Its super-sensitive reception brings you AM, International Shortwave, Longwave and full-carrier AM Amateur Broadcasts, Ship-to-Shore and Ship-to-Ship, Marine and FAA Weather Services. And glorious FM, too! See the famous 9-band *Trans-Oceanic®*, Model Royal 3000-1, at your Zenith dealer's!

ZENITH ®

The quality goes in before the name goes on

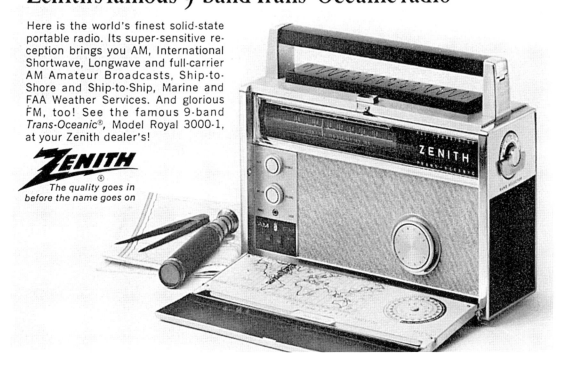

Royal 7000 Series Trans-Oceanic.

Royal 7000 Series Trans-Oceanic.   Royal 3000 Series and Royal 7000 Series Trans-Oceanics.

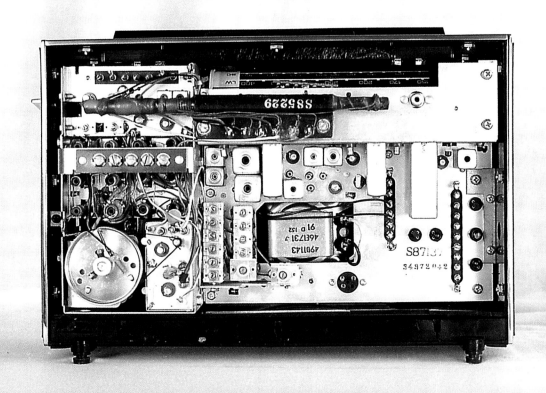

Royal 7000 Series Trans-Oceanic.

ROYAL 7000 SERIES
A Brand New Radio!

Introduction

By the late 1960s, the design of the Royal 3000 had become dated, both electrically and aesthetically. Zenith determined to design an all-new Trans-Oceanic with modern circuitry and an all-new cabinet; the resulting new Royal 7000 Trans-Oceanic was introduced as the Royal 7000Y with a press release on April 29, 1969.

The new Royal 7000 featured 11 bands and was "powered to span the globe and receive even more information locally." One addition was a fixed crystal VHF weather band at 162.55 MHz. The set was tuned to the "weather position" to hear up-to-the-minute local weather reports from the National Weather Service. The second band addition was shortwave: a division of the previous 2-9 MHz band into one band spanning 1.6 to 3.5 MHz and a second spanning 3.5 to 9.0 MHz. This new dial division, then, made a total of 7 shortwave bands, in addition to long wave, FM, AM, and the fixed frequency weather band. Other improvements in the Royal 7000 included a beat frequency oscillator (BFO) and manual RF gain controls; these latter two controls enhanced the reception of Morse code (CW) signals and the burgeoning new voice transmission mode - single side band (SSB).

Versions

ROYAL 7000Y

Introduced
May 1969 for the 1969 Line.

Last Offered
1970 Line.

Chassis
18ZT40Z3.

ROYAL 7000Y-1

Introduced
December 1970 for the 1971 Line.

Last Offered
1972 Line.

Chassis
18ZT40Z.
In the 1971 Line, the weather band crystals were replaceable by the owner. That changeability is the major difference between the Royal 7000Y and the Royal 7000Y-1.

ROYAL D7000Y

Introduced
December 1972 for the 1973 Line.

Last Offered
1978 Line.

Chassis
500MDR70.
The Royal D7000Y introduced a **tuneable** weather band and contained some minor cosmetic changes (refer to the chapter on Parts Commonality for a discussion of these subtle differences).

Cabinet Design

The industrial designers of the Royal 7000 obviously were well aware of each of the major weak points of the original transistorized Trans-Oceanic cabinet design. These included:

1) The almost disastrous metal plating problems causing long-term pitting and massive blistering of both polished and brushed chrome parts.
2) The overly fragile main carrying handle caused by combining it with the Waverod antenna.
3) The inclusion of the battery box inside the primary cabinet led to, at best, a corroded interior cabinet and, at worst, destruction of many of the delicate electronic components of the receiver.

4) Expeditions which evaluated the Royal 1000/3000, as well as a number of more "normal" customers, had complained of the difficulty of moving the supposedly portable Trans-Oceanic while listening to it. To simply move the Royal 1000/3000 across the room or tent, it was necessary to close the front door, collapse the Waverod and relock the carrying handle.

The new cabinet design of the Royal 7000 was a magnificent example of "user-friendly" ergonomically correct industrial design. The physical design of the radio allowed it to be what it was designed and purchased to be: a superb portable radio which would be as at home in the drawing room as it was on expedition in the high Himalayas. The designers attacked the problems of the earlier transistorized Trans-Oceanics at their roots.

The fragile carrying handle of the Royal 1000/3000 radio was supplanted with a large collapsible and *very* sturdy carrying handle. The materials problems, particularly those with brushed and polished chrome, were also addressed. The earliest Royal 7000s are now over 20 years old; the materials seem to be standing the test of time extraordinarily well, with no known examples of the pitting and blistering which plagued the Royal 1000/3000 Series.

The problem of battery leakage to the interior of the receiver was solved by providing a double walled removable rear cabinet wall. When the first smaller rear cabinet wall is removed, a battery compartment is revealed. This compartment is comparatively well-sealed from the interior of the receiver, all but eliminating the problems associated with battery leakage. The inner face of the removable battery compartment cover also provides a convenient storage compartment for a small accessory earphone and the detachable AC power cord.

The final, larger, rear wall of the cabinet could be removed by unfastening three screws which allowed removal of the entire rear cabinet wall to expose the chassis and speaker. Zenith engineers also eliminated the 12v DC exterior input port and its attendant and easily misplaceable outboard AC power supply cube. The Royal 7000 was equipped with a complete 115/230v AC power supply which was built into the removable rear cabinet wall. Switching between batteries and house current was done automatically when the AC power cord was attached to the rear of the receiver.

The fragile carrying handle and the in-use moving difficulty of the Royal 1000/3000 were equally well-addressed by the designers of the new Royal 7000. To accomplish this, the Waverod was removed from the carrying handle and placed at the very upper rear of the cabinet. The Waverod could be deployed from this location whether the cabinet doors were open or closed. The two-part hinged cabinet door of the older Trans-Oceanics was replaced with two separate horizontally hinged front doors. The lower of these doors swung down to reveal the lower 2/3 of the receiver's front panel. The inner surface of this door contained a large azimuthal compass wheel to assist in direction finding. After the lower door is horizontal, it could be made to slide back into the radio cabinet and largely disappear from view; it could remain in that position while the receiver was carried from place to place.

The upper door was equally ingenious. Much like the old tube-type Trans-Oceanic front door, it was hinged halfway back on the top and rotated upward to reveal the dial of the Royal 7000. This door cleared the main carrying handle when the handle was in its collapsed position. When this upper door was open, its inner (now front) surface revealed a time zone-oriented world map and a "time strip" at the lower edge of the map. This time strip could be moved laterally to align the correct time under each time zone. This is by far the most useful arrangement for world time conversion the authors have seen.

Finally, the designers of the Royal 7000 created a very beautiful radio with a definite 1970s look. Great areas of bright and brush chrome were almost perfectly balanced with large areas of tough, high-impact black Cyolac plastic. The material and color mix was highlighted by the colorful back-lit dial and a light blue rendition of the world map. All in all, the designers of the Royal 7000 created a radio that surely would have been enthusiastically supported by Commander McDonald and his crew of "real radio men."

Electronic Design

Two major weaknesses had been noted in the electronic design of the AM-FM Royal 3000 Trans-Oceanic which needed correction in a new model:

1) The lack of a Beat Frequency Oscillator and a manual RF Gain control made reception of the now-popular SSB mode of voice transmission extremely difficult (there was a BFO kit available for the Royal 3000; refer to section on Accessories).

2) International shortwave broadcasting had undergone explosive growth in number of stations and in transmitter power during the 1960s, causing major overcrowding on many of the International Shortwave Broadcast bands. Additional receiver selectivity, now designed into the Royal 7000, was required to combat this problem.

Further, the basic circuit of the Royal 3000 was actually that of the Royal 1000 which had been designed in 1957, and virtually every aspect of the electronic design of the Trans-Oceanic was enhanced by Zenith engineers for the Royal 7000. The twelve-transistor Royal 3000 was replaced by the 18-transistor Royal 7000 chassis. Three of these additional transistors were allocated to the tuning section of the new weather band. One was added as a voltage regulator and one used for the new Beat Frequency Oscillator for tuning SSB and CW signals. The final new transistor was allocated to a more powerful audio section.

For the very first time in the Trans-Oceanic line, Zenith design engineers also provided switchable band-width control. The set was operated normally in the "Wide" position to achieve the best audio fidelity; in crowded band conditions, it was possible to switch to a "Narrow" position and thereby eliminate much of the interference.

Zenith made one other major design and manufacturing decision of note in the Royal 7000: they decided to maintain a heavy metal chassis and traditional hand-soldered point-to-point wiring. That decision was a commitment to traditional American views of quality, but it locked the long-lived Royal 7000 Series into a very labor-intensive manufacturing process. This also made it very difficult for the Royal 7000 to compete with the emerging, highly automated production of East Asia.

Price

The price of the Royal 7000Y at introduction was $270. The final retail price for the Royal 7000 Series, in 1979, was approximately $300.

Selected Print Advertising

The introductory advertisements for the Royal 7000 were published as front advertisements in *National Geographic*, November and December 1969. The only other mass media print advertising was a beautiful, full page ad in the 1970 *World Radio and TV Handbook*. This advertising campaign can only be characterized as scant. Although it is true that by this time the Zenith corporate flagship was color television, one wonders at the logic of investing corporate research and development funds in creating an entirely new Trans-Oceanic and then, seemingly, condemning it to a lingering death through almost a total lack of advertising.

Introducing Zenith's all-new 11 band Trans-Oceanic®

The last word in radio performance.
The biggest word in radio sales.

Zenith announces the fabulous new Royal 7000Y ... the finest *Trans-Oceanic* ever...with powerful performance and 11-band coverage for a whole new world of listening pleasure.

Eleven different radio wave bands offer enormous versatility in program reception. Included are: new crystal control VHF weather band for instant monitoring of the continuous U.S. weather reports. FM—with drift-free AFC. Longwave weather and navigation band. Standard AM broadcast band. Two marine, weather, ship-to-ship and amateur shortwave bands. And, with *five* International shortwave bands, it's powered to tune in the world.

Band spread tuning on International SW bands electrically spreads each station up to 10 times farther apart on the band ...makes tuning easier and faster, with pinpoint accuracy.

Other great new quality features include Zenith exclusive Beat Frequency Oscillator for outstanding receptic of SSB, CW, Consol and Consolan transmissions. Earphor (included) for the handy, front-mounted jack. "Norm-Shar IF switch. Manual RF gain control. Tuning and Battery Lev Meter for visual as well as audible tuning, and for reading th battery voltage. Tilt-out front panel light—as well as a dial ligh Built-in power supply for either 115V or 230V AC power.

Zenith's all-new Royal 7000Y *Trans-Oceanic* offers th finest features and multi-band performance in the radio indust today. Be sure you have it—to roll up big new sales in '69!

Why not sell the best **ZENITH**®

*The quality goes in
before the name goes on*

The Royal 7000 was a welcome guest at this Hawaiian luau shown in a Zenith publicity photograph from 1969. *Courtesy of Zenith.*

"Direct from La Scala...the opening performance of Pagliacci." 25 meter band

"Wave heights expected: 15 to 18 feet off Montauk Point." 162.55 MHz weather band

"The British are 100 for 5 at cricket in Karachi." 19 meter band

"This is North American Service of Radio Moscow with the news." 16 meter band

"Boise: altimeter, two niner seven niner. Ceiling, twenty-five hundred." 150-400 KHz band

"This is ZD8ZAD Ascension Island calling W9DCN Peoria." 13 meter band

"We've taken two marlin here at Pompano...looks good." 1.6-3.5 MHz band

"Next tone begins at 14 hours, 6 minutes, Greenwich Mean Time." 31 meter band

"Hurricane winds of Force 12 expected from 20 hundred hours." 3.5-9 MHz band

"Bulletin: the Coast Guard has sighted survivors." Standard Broadcast band

"And now WEFM presents West Side Story." FM Broadcast band

Presenting the Zenith 11 band Trans-Oceanic.
The last word in radio. About $270.*

Built to tune in the world with simplified precision tuning. Electronic band spread tuning widens dial space for easier station selection on five shortwave bands. BFO control for Single Side Band and CW code reception. Tuning meter shows signal peaks for fine tuning. Manual RF gain control to pinpoint RDF locations and facilitate SSB and CW reception. *Model Royal 7000Y-1, price optional with dealer.

ZENITH

The quality goes in before the name goes on

The Royal 7000 goes full circle in this publicity photo taken with MacMillan's exploration vessel, the *Bowdoin*, which was used on both the 1923 and 1925 Arctic expeditions. *Courtesy of Zenith.*

R-7000 Series Trans-Oceanic.

R-7000 Series Trans-Oceanic.

Royal 7000 Series and R-7000 Trans-Oceanics.

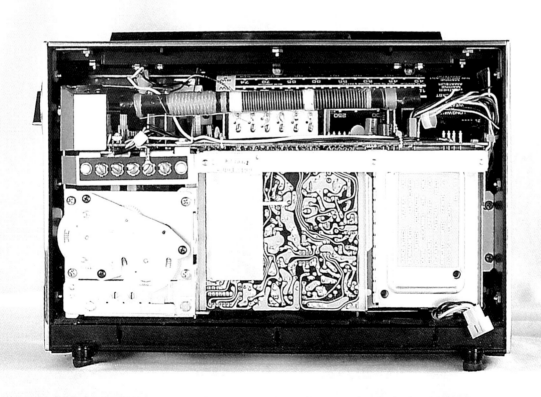

R-7000 Series Trans-Oceanic.

R-7000
The Last Trans-Oceanic

Introduction

The last model of the Zenith Trans-Oceanic was introduced in a news release published on May 25, 1979. The R-7000 was a brand new receiver, representing an entirely new generation of electronic products: integrated circuits rather than transistors, printed circuit boards rather than point-to-point wiring. This new technology was placed in the old cabinet of the previous model, the Royal 7000, and used the old model number, 7000, dropping the term "Royal" from its title. This series of decisions seems totally wrong-headed and confusing from the marketing point of view. These decisions alone should have doomed the set to fail miserably in the marketplace: it did. The R-7000 was designed in Chicago, and for the first year, was manufactured at Zenith's Chicago Plant #2; as such, it was the last portable radio manufactured in the United States. The second and third year of production took place at a Zenith assembly plant near Taipei, Taiwan, using American-made parts.

Versions

The R-7000 was produced in three (yearly) versions, known as the R-7000, the R-7000-1 and the R-7000-2. From the point of view of longevity and ease of use, it is preferable to collect the R-7000-2, which is believed to be the only version with the chassis number 2WMR70. The earlier two versions (chassis 2WKR70) were plagued with a faulty tuner drive belt. The belt had far too much "backlash" and flexibility to make shortwave tuning reliable. There were a number of other technical problems with the 2WRK70 chassis, as well; prudence alone would suggest obtaining the later 2WMR70 chassis, if at all possible. A white paper adhesive tag showing chassis number and model was affixed to the outer surface of the cabinet bottom and should be checked prior to purchase.

Number Produced

Unlike any previous Trans-Oceanic, the R-7000 was designed and produced as a "specialty item." There were only 25,000 units produced annually. Twenty-five thousand units was, of course, far below the annual production of other models and would at least partly account for the relative rarity of the R-7000 today.[8]

Last Offered

The last Zenith Trans-Oceanic was offered for sale late in the 1981 model year.

Chassis

2WKR70 (bad) and 2WMR70 (good).

Cabinet Design

The basic cabinet of the R-7000 was identical to the previous model, the Royal 7000. However, the "face lift" design done by Rick Althans of Boldt and Associates, produced a singularly striking radio.[9] The major styling changes from the Royal 7000 Series radios were that the front panel and the speaker grille, the inside of the front door, and the world map/time chart were all predominantly black rather than the chrome/silver of the Royal 7000 Series. The smaller controls were also rearranged on the left-front, and the Royal 7000s single small meter on the left was replaced by two rather larger meters, which were moved up to the right-hand end of the dial glass. A new two-part tuning control knob was developed for the R-7000. It had a concentric smaller fine tuning knob which tuned at a 40 to 1 ratio, accomplished by mechanical rather than electrical means.

Electrical Design

From the point of view of the entire forty-year history of the Zenith Trans-Oceanic line, the electrical design of the R-7000 represented two rather radical design departures: adopting a "general coverage" rather than isolated electrical bandspread philosophy for the shortwave bands, and adopting state of the art modular design, based on integrated circuit chips and military-grade glass printed circuit boards.

Frequency Coverage

From a user's point of view, the R-7000 marks a complete departure from all other Trans-Oceanics. Previously, the shortwave tuning coverage of Trans-Oceanics had been largely "limited" to the International Broadcast bands. Each tuning range was, in fact, an "electrical bandspread" of the major International Broadcast bands. In older Trans-Oceanics, the only general SW coverage was limited to two bands: 2 to 4 MHz and 4 to 8 MHz. These represent the lower 1/3 of shortwave frequencies and were particularly useful for mariners and travelers in the Third World. The new R-7000 abandoned this "focused" approach to frequency coverage in favor of six separate dial bands which continuously covered the entire SW spectrum, from 1.8 kHz to 30 MHz. The difference in these two approaches can be illustrated using the popular 31 Meter International Broadcast band, which runs from about 9.4 to 10.1 MHz. On the Royal 7000 Series, this band was given its own dial, "31 Meters"; the new R-7000 covered 31 Meters in the middle of the "SW 3" band. The same 9.4 to 10.1 MHz segment, which occupied 3.5 inches in the previous receiver, measured barely one inch on the R-7000. It is also fair to note that the new continuous frequency coverage might appeal to a wider market with a wider constellation of interests. This "wider market" approach was also evidenced by the inclusion of a separate band for "personal communications"; this is, of course the now-notorious Citizen's Band.

The "wider market" philosophy also led designers to add a second tuneable VHF (high) band to be used to monitor the public service bands. The adoption of this second VHF band was disastrously timed, since truly automated VHF scanners were then beginning to dominate that market; the CB craze was also winding down. These decisions proved to be very poor timing.

Other Electrical Design Decisions

In an effort to compete in the middle price bracket, Zenith executives and the designers of the R-7000 delivered the unkindest blow of all: they chose not to incorporate a digital read-out dial and a digital (keypad) frequency input. These capabilities had been on the market for at least five years in Japanese-produced receivers designed for the radio amateur and the upper end of the general listening markets.

In 1981, only two years after the R-7000 was introduced, SONY of Japan introduced the ground breaking ICF-2001, also targeted at the middle price range. The major features of the now famous 2001 were digital frequency read-out and numerical keypad frequency entry tuning. The 2001 was priced very competitively with the R-7000, was 1/4 its size and weight and could be tuned reliably by a first time user--*there was no contest*!

Price

At introduction, the price of the R-7000 was $379.95. In the ignominious end the R-7000 was discounted far below dealers cost (to as little as $50) by many local dealers and discounters simply to clear their shelves.

Selected Print Advertising

There are no known general circulation magazine advertisements of the R-7000.

TRANS OCEANIC R-7000
A masterpiece of
innovative technology.

Superior technological achievements enabled our engineers to design the Trans Oceanic R-7000 to rigid commercial specifications, resulting in greatly improved performance, stable, reliable reception and, in effect, 12 radios neatly fitted into one classically styled compact unit.

Many of today's advances in electronic technology were brought about by requirements of the armed forces and financed by government at research cost far beyond the range of a consumer products manufacturer. Some of the most useful of these have been incorporated into the Trans Oceanic along with sophisticated engineering improvements devised in Zenith's own laboratories.

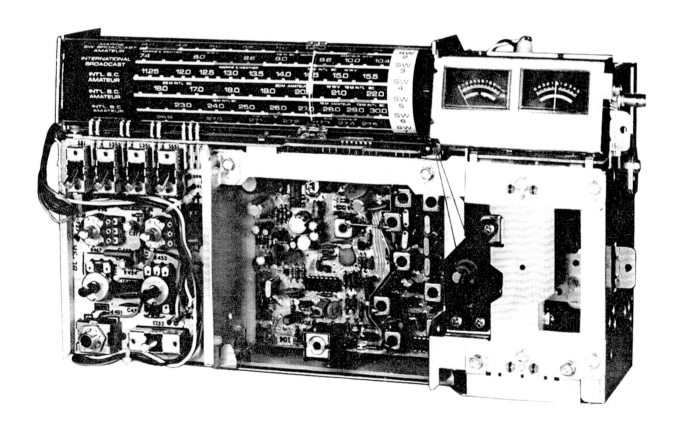

TRANS OCEANIC R-7000

Chapter 8

TRANS-OCEANIC
ACCESSORIES

Very few accessories existed during the forty-year life of the Trans-Oceanic series. Those that were offered dealt almost exclusively with further equipping the Trans-Oceanic for use abroad or in the field. These were always offered at extra cost and most were probably quite rare even when new. When Zenith left the radio business in 1982, Trans-Oceanic Part Production and Inventory were closed soon after. In hopes of finding some few remaining parts or accessories, the authors submitted a rather extensive list of part numbers to Zenith corporate headquarters. A computer check of Zenith inventories was performed; and, most unfortunately, no remaining parts or accessories were discovered. Any remaining parts or accessories must, therefore, reside in the hands of a few of the older Zenith dealers or with major Trans-Oceanic collectors. This listing is provided for interest, completeness and to assist those lucky enthusiasts who unearth a few of these very rare items.

The Tube Years

The first accessory offered for a Trans-Oceanic was the "Hush-a-Phone" pillow speaker. It was offered as an accessory to the 7G605 Trans-Oceanic Clipper.

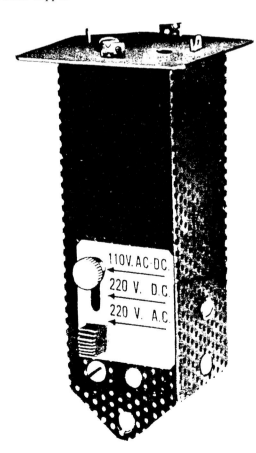

Power Supply (or Ballast) Adapter. Zenith Part # S15715.

The most common accessory of the tube era was a "**Power Supply (or Ballast) Adapter**" module which allowed the Trans-Oceanic to be used in countries where 220v was the normal household current. The military version of this adapter, used with the R-520/URR, carried the Part Number Z-1. The civilian unit was Zenith Part Number S-15715. This adapter unit measured 1.5" x 1.5" x 3.75" and plugged into the right front of the chassis top (when viewed from the rear). The module contained a dropping resistor and a few other components to lower 220v to the 120v that the remainder of the power supply normally received. The Ballast Adapter was available on all miniature tubed chassis, beginning with the G500. When the Adapter is not plugged into its chassis port, the port is covered by a small "L" shaped metal plate.

The only other known tube era "accessory" was a **spare tube storage strip** and is found in some G500, H500 and 600 Series radios. This metal strip was usually mounted on the inside surface of the rear cabinet door and contained five metal clips so that an entire set of spare tubes could be carried right in the receiver. No advertising has been found which indicated that this clip-strip was offered for extra cost. No records were found which could explain its presence or absence on a particular Trans-Oceanic.

The Solid-State Years

A number of accessories are known to have existed, at times, for transistorized Trans-Oceanics. These were:

Royal 1000 Series

A) A matching Zenith **Earphone Kit**, Part Number B39-24, was offered during most of the 1000 Series years. This was a single "mini-" earphone and came in a plastic box. The earphone, in the box, was then to be carried under the elastic cord found on the lower right-hand side of the chassis top (inside the rear of the cabinet).

B) A **BFO Kit**, Zenith Part Number S-53472, designed to permit the reception of "amateur (CW) transmissions" was also "optional at extra cost" in 1962 and later years.

C) An **Antenna Transformer Kit**, Part Number S-49339, was offered in 1964: "Helps increase Broadcast Band reception in fringe areas."

D) The introduction of the "-1" models (such as the Royal 1000-1) for the 1964 line also required an "**External AC Power Supply**, Part Number S-64352, available at extra cost. The "battery eliminator" type power supply plugged into a built-in jack on the side of radio and into the AC outlet. It saved battery drain by permitting the use of household current where available. Zenith Part Number S-65074 was a similar adapter available for areas having 230/115v as the normal household current.

Royal 3000 Series

The same four accessories available to Royal 1000 Series owners were also available to owners of the new Royal 3000 Series receivers. There were also a number of accessories added to the line with the introduction of the Royal 3000 for the 1964 line. The total accessory line for these two models was:

A) **Zenith Earphone Kit**, Part Number B39-24 (see above).

B) **BFO Kit**, Zenith Part Number S-53472 (see above).

C) **Antenna Transformer Kit**, Part Number S-49339 (see above).

D) **External AC Power Supply**, Part Number S-64352 (120v AC) and Part Number S-65074 for 220/115v (see above).

E) A removable ferrite bar **medium wave Wavemagnet** offered as an accessory during the Royal 3000 years, Zenith Part Number S-42212. From photographs, this accessory seems to have been identical to the Wavemagnets which were supplied with the Royal 1000 Series. It is likely that these were not supplied as original equipment in the Royal 3000s because the in-board storage space available in the Royal 1000 Series was totally taken up with the new FM sub-chassis.

F) **Adapter Kit**, Part Number S-51194, "lets you enjoy Royal 3000 through speaker system of some high fidelity or stereo instruments."

G) **Standard Low Impedance Double Headphones**, Part Number 39-34.

Royal 7000 Series

When the Royal 7000 was introduced in April 1969, press releases listed three "optional extra-cost accessories: swivel base for more positive direction finding; mobile plug-in antenna; headphones for private listening." Since the new Royal 7000 Series provided both an internal BFO and an onboard 120v AC power supply, these two items were, naturally, not listed as 7000 Series accessories. Early PR releases do show a mini-earphone, now boxless, tucked under the elastic cord in the power cord storage compartment. This is probably the mini-phone previously listed as Part Number B39-24. Only the earliest PR also mentions a "mobile plug-in antenna," presumably the MW Wavemagnet, Part Number S-42212, as above. No mention is made of an adapter kit (item F above) to allow the Royal 7000 Series receiver to act as a tuner with a stereo system. As with the 3000 Series, the Royal 7000 receivers provided such an output jack on the rear of the receiver.

The most striking accessories ever offered with the solid-state Trans-Oceanics were two items introduced at the very beginning of the Royal 7000 Series:

A) A padded vinyl **Carry Case** was offered during, at least, the early Royal 7000 years. It was announced to dealers in December 1969 in internal Form R8324-1268: "This high quality carry case is completely padded with foam rubber to provide extra protection for the Royal 7000. The case is made out of durable Naugahyde in a striking black color. Functional zippered pouch designed to hold radio log charts and handbooks. The front of the carry case drops down to permit playing radio in case." The internal memo went on to advise dealers to "Order from your Zenith distributor as Form Number R8318."

B) A rotating, swiveling stand to hold the Trans-Oceanic when it was used for long wave or medium wave direction finding was offered throughout production of both the Royal 7000 Series and the R-7000 Series. It was introduced with the following commentary: "**Swivel Base Platform** - The feet on the Royal 7000 are grooved to lock into a rotatable platform that can be purchased as an accessory. Set can be securely fastened to a table or similar object during rolling sea or wherever rugged conditions are encountered."

No Zenith part number had been found for this base until similar descriptions were published for a "Swivel Base Platform" to accompany

Royal 7000 mounted on Swivel Base accessory.

the R-7000 Series receivers. Since the bottom of the two receivers is identical, a fairly safe assumption is that the accessory base was identical. The part number for the R-7000 swiveling base was S-75893.

R-7000 Series

Three accessories were associated with the R-7000 Series receivers: the previously mentioned swiveling base platform, a 12v Auto Adapter Cord and a new Zenith dual headphone set. The headphones were not mentioned in brochures as an "official" Trans-Oceanic accessory, but were often shown and called out by model number in R-7000 Series PR photos. Quoting in detail:

A) "**Swivel Base** - A rotatable platform which can be fastened to a table or other stationary object to secure the Trans-Oceanic in place (especially useful on ships). Zenith Part Number S -75893."

B) **12v Auto Adapter Cord** - Zenith Part Number 11-328.

C) **Optional Stereo Headphone** - Model 839-56, "featuring 'open' design, and rotary tone and volume control on each earpiece." [Latter from cut line on back of 1979 PR photo.]

No accessible records exist to ascertain how many of each item were sold during the production years. All Trans-Oceanic accessories, whether for tube or transistor models, are quite rare today.

Note:

All of the above information was developed from Zenith brochures found in the annual "spec books" or from original Public Relations photos found in the Zenith archives.

Chapter 9

COLLECTING TRANS-OCEANICS
Finding, Buying and
Collecting Trans-Oceanics

Because of the forty-year run of Trans-Oceanic models, several generations experienced the joy of owning this fine radio when it was new. Today, as many people nostalgically connect to their past, some have considered acquiring another Trans-Oceanic. Other contemporary purchasers know little about Trans-Oceanics, but are impressed enough to want to know more. No matter how the owner and radio come to know one another, the relationship often leads to the desire to own other Trans-Oceanic models and a collection is often the result.

Although perhaps relatively new to you, collecting Trans-Oceanics is not a new hobby. Vintage radio collectors have dealt in them since the 1960s and radio hobby journals have carried articles and advertisements for used Trans-Oceanics for many years. In recent years, other collectors have appeared: 1940s, 1950s and 1960s nostalgia collectors; industrial designers looking for perfect examples of the designer's art; a growing group of people who first heard shortwave on a Trans-Oceanic and now have the money to buy one; a larger number of knowledgeable antique dealers buying for the upper-class trade; and, speculators who plan to put them away until they are more valuable. The collecting market is definitely growing, the radios are becoming scarcer, the price is going up and in order to be able to obtain the models you need, you have to work harder and collect smarter.

Should You Buy One At All?

If you do not yet own your first Trans-Oceanic, you have to decide if you want to become a custodian of even one. There are many advantages, of course, to owning even a single Trans-Oceanic. Thankfully, at least in the mid-1990s, ownership is still relatively inexpensive. If you have decided to take "the plunge," you still have many choices to make. Probably the first is to decide whether to hunt for one of the tube models or one of the transistorized sets. There are several issues to consider: solid-state radios drove vacuum tube radios off the market for some very good reasons. Probably the most important of these is reliability. Vacuum tube receivers require more maintenance and fail more regularly than do transistorized models. If you are adept at "hollow state" electronics, you need not worry. If you are an electronic neophyte but still love the glow of vacuum tubes, you should be aware that it is still possible to get tube receivers serviced through the vintage radio hobby community. Most beginners worry about tube availability--don't! Tubes are available at reasonable prices from many sources (refer to Appendix II).

If the modernity of transistors seems to be for you, or if the look of the late 1950s and 1960s is appealing, you should purchase transistorized Trans-Oceanics. In addition to their reliability, they take up less space. Furthermore, it is still sometimes possible to get transistorized Trans-Oceanics serviced at Zenith dealers.

In any case, there is no need to agonize over the decision too long. If you change your mind, or want to get out of Trans-Oceanic collecting completely, it is easy to do so since there is a very active market for Trans-Oceanics. You should be able to sell yours for at least what you paid.

Some Thoughts About Collecting

Whether you are just contemplating your first Trans-Oceanic or you already own several, the chief rule of collecting is organization. The first step in organization is to decide just what it is you are collecting. Would an all tube-type (or all transistor-type) collection satisfy your need for collecting, or do you want all ten models? Do you want just a representa-

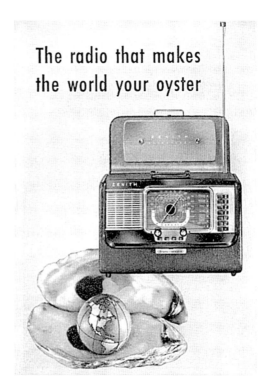

tive of a model line, such as any Royal 1000, or do you want all representatives: a Royal 1000, a Royal 1000-1 and a Royal 1000-D? Are you also interested in collecting the ephemera, or "paper," that is part of the radio: owner's manuals, handle tags and advertisements? It is important to decide the bounds of your collecting activity, since sources are often different.

Another part of organization is record keeping. Start a card file or a simple data base on your collection. At the least, it should reflect date and place of purchase, amount paid, general condition, serial number and any other information of importance. As you see other radios, you can expand your notes and thus educate yourself on the "going cost" of other receivers in the same condition as yours. These notes are also beneficial for resale and insurance purposes.

Sources

The number of sources for obtaining Trans-Oceanics is limited only by the collector's imagination and degree of experience. The following list indicates the sources most often used by collectors and sellers and hopefully should help you with your search:

Older friends and relatives
Estate sales
Garage, church and yard sales
Classified and "Trading Post" advertisements (either advertisements that you find or ones that you place)
Pawn shops (although Trans-Oceanics are not as common there as they were once. They are considered "old technology" and therefore not worth much on pawn)
Thrift stores
Ham fests and amateur radio flea markets
Antique radio events and meetings
Vintage radio hobby publications (addresses in Appendix II):
Antique Radio Classified
Ham Trader Yellow Sheets
Wireless Trader
Transistor Network
Electric Radio
Long established radio repair stores in older parts of town
Radio equipment dealers that accept trade-ins on used equipment
Antique stores (although if found there, the price usually will be high)

Price

Expensive Trans-Oceanics were not casual purchases when new. Consequently, many of the 1,500,000 or so that were made have been lovingly stored away and are now coming on the market, primarily through garage and estate sales. Finding radios in garage and estate sales is the least expensive but most time-consuming approach. Although prices vary widely, Trans-Oceanics from these informal sources have tended to cluster in the $40-$50 range in the early 1990s. Buying Trans-Oceanics within the vintage radio community is quick and easy, and the condition of the radio is more readily determined. Prices, however, are about double that found on the informal market. In general, common model Trans-Oceanics have been selling from $90-$150 in the vintage radio community in recent years. Bunis' *A Collector's Guide to Antique Radio* is a readily avai able and outstanding price guide which covers the prices of most vintage radios and is an excellent adjunct to specialized Trans-Oceanic collecting.

Purchasing Trans-Oceanics from general antique dealers is usually an expensive option. Although there are exceptions, most radios in antique stores are priced at double their value within the vintage radio community.

Condition

As you start collecting, you will soon discover that Trans-Oceanics are found in a full range of conditions and that price and condition parallel each other closely. At the bottom of the condition spectrum is the "parts radio," a Trans-Oceanic that has seen better days and is now ready to be thrown out. It may be musty and water damaged, have missing parts or just be badly battered. Often, parts radios should not be overlooked. Virtually every one will have some usable parts--knobs, screws,

face plates, a Wavemagnet, a handle--which may be all that is necessary to turn an almost perfect future find into a classic. Parts radios are usually cheap. It is a good idea to have at least one parts radio for every model in your collection. Even a small screw, no longer available from Zenith, may be enough to pay for the radio.

In the middle of the condition spectrum are found a large number of "fair" to "good" Trans-Oceanics. The "fair" and "good" radios look their age and may have a few dings and missing pieces. They may or may not work, but they are restorable. Prices vary most widely with these average radios and a collector has to decide how valuable they will be to his collection.

Occasionally, a collector will encounter a "mint" Trans-Oceanic. Such a radio will most likely have been safely stored for many years and will take only a once-over with a cleaning cloth to restore it to its original beauty. Such a Trans-Oceanic is a rare find and will almost always command a high price.

The condition that a collector seeks for his collectibles is a matter of personal preference. Some people build a collection of Trans-Oceanics that look like each just came out of the box, while other people choose to have an average collection. For very different reasons, each of these collectors have exactly what they want in a collection. As you become more experienced, you will discover that there are far more "fair" to "good" Trans-Oceanics than there are "mint" ones; to avoid frustration you might want to at least start with those in average condition. Many average Trans-Oceanics can be led well toward mint condition by employing the techniques contained in the chapters on restoration in this book. In addition to having an attractive and useful radio, there is a great sense of pride associated with salvaging one of these beautiful radios.

Whether a Trans-Oceanic works is another aspect of condition that must be carefully considered. The reasons for a radio not working can be many; often the fix is simple; other times it is not. If you know absolutely nothing about electronics, you should think carefully before you purchase a nonworking radio; it may be costly to pay someone to repair it. However, any nonworking radio in a good to excellent case, at the right price, should be purchased and saved for the time when you find a working one in a bad case. Nonetheless, consider it and your own ability carefully before you spend your collecting dollars.

Tips

Certain areas of the country seem to be better suited for finding Trans-Oceanics than others. Since Trans-Oceanics were very expensive when new, they were owned mostly by those wealthy enough to purchase them, chiefly professionals. As these people aged and settled in retirement communities, their Trans-Oceanics often moved with them. The right number of years have now passed so that these people are either selling their extra things as they move to smaller homes or their belongings are being sold by their heirs. As a result, springtime garage sales in areas heavily populated with older retired professionals often offer better chances to acquire good Trans-Oceanics than do other areas.

When you locate a Trans-Oceanic, look it over carefully for damage and missing parts. Remember, the parts are no longer available from Zenith and if anything is missing or badly damaged, it can only be replaced with parts from a parts radio. Although not necessarily grounds for rejection, missing parts may provide enough frustration to dampen some of the enjoyment of your hobby.

There are several common flaws found in Trans-Oceanics; you should examine each radio thoroughly to determine if these flaws are present and of concern to you:

Tube Models
Bent or broken Waverod antenna. These are difficult to straighten; if broken, they will require a replacement from a parts radio.
Missing top lock catch on the Waverod antenna. If gone, the antenna either will not collapse all the way or else will collapse so far into the case that it will be hard, or impossible, to withdraw. The missing latch does not affect the operation of the radio.
Broken or missing "bale" on the main cabinet latch. These are potmetal and cannot be repaired.
Broken hinge, usually on the right side, on the drop down chart compartment in the 600 Series. This is a fairly common flaw and often is not repairable, you just have to learn to live with it.

Unglued Black Stag material, a result of storing in damp areas. This is not a major flaw (although it might look that way) and is easily corrected (see chapters dealing with restoration).

Transistor Models

Broken or split handle in the Royal 1000 and Royal 3000 models. The handle is hollow to hold the Waverod antenna and is a major weak point. If the handle is not in good condition, the radio cannot be carried properly and you run the risk of dropping it. If the handle is not badly cracked, it may be disassembled and repaired/reinforced from the inside.

Missing battery box in the Royal 1000 and Royal 3000 models. Unless modified (see Electronic Restoration-Solid-State section), these models run only on batteries; if the battery box is missing, you will not be able to use the radio. These boxes are sometimes found, but they are rare. It is also possible to wire a battery eliminator directly to the battery input plug on the chassis (see Electronic Restoration-Solid-State). If the battery box is missing in either the Royal 1000-1 or Royal 3000-1, the problem is not as serious since they have a built-in jack for 110v use (see Electronic Restoration-Solid-State section for polarity of adaptor).

Chrome pitting and puckering on the Royal 1000 and Royal 3000. This is a very common malady on these models and is considered nonrepairable using practical current technology.

All indications suggest that more people are collecting Trans-Oceanics now than ever before. For some, the hunt is more fun than the catch; for others, the Trans-Oceanic in the den, delivering news from the BBC, is a tie to another time and worth any cost. There also seems to be a market developing of collectors acquiring Trans-Oceanics for investment. As with any form of collecting, the collector finds reward for reasons best known to him/herself. Happy Hunting!

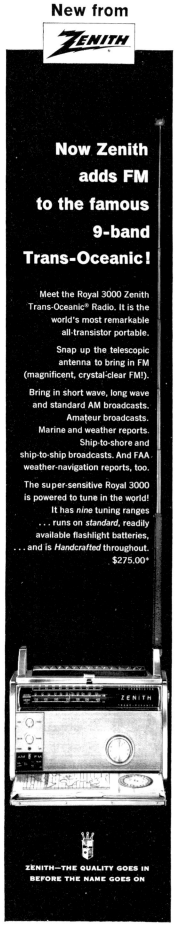

128

Chapter 10

PHYSICAL AND ELECTRONIC RESTORATION OF TUBE MODELS

General Overview

Restoring any historical artifact to useful life is a very rewarding endeavor—literally bringing something back from the grave. It is a pursuit which gives participants many hours of pleasure, no matter whether they are professional restorers or amateurs. There are really three parts to any radio restoration: 1) deciding what to do and how to do it; 2) physical restoration; 3) electronic restoration. Careful study of all three activities which are covered in this chapter should provide a road map to successful restoration of tube-type Trans-Oceanics.

The Zen of Trans-Oceanic Restoration

There will never be any more Zenith Trans-Oceanics in the world than there are right now. That means that each of us who owns a Trans-Oceanic, no matter how beaten and battered, is in a very real sense only a custodian of a marvelous radio. This "non-replaceable" aspect to Trans-Oceanic ownership also adds a dimension of ethics and philosophy to every decision concerning the repair and restoration of any Trans-Oceanic. *YOU* must decide what to do.

The first issue to ponder is just **how much** restoration to attempt. There are two schools of thought: 1) **do as little restoration as possible**: cleaning, minor repair and little more; after all these items are old and should look it! 2) **bring the item back to as nearly "showroom" condition as possible**: a brand new radio is what the original purchaser wanted and got! Most museum curators lean toward the minimal restoration approach. Most radio enthusiasts fall between the two philosophies but lean toward the "showroom" approach.

If you are new to radio restoration, it makes good sense to begin your career as a "minimalist" restorer. It is often very difficult or impossible to undo/repair damage caused by a misinformed or hasty restorer. Tragically, people ruin radios this way.

The very best approach for a novice Trans-Oceanic restorer is to plan to revive the radio in a number of stages. The first stage would be to clean the radio and to do minimal safety and operational electronic repair. The second and maybe a third stage of cosmetic restoration, discussed later, would be completed before attempting a full cosmetic restoration.

If you are a neophyte at radio restoration, one of the best suggestions would be to obtain a "companion" Zenith Universal receiver and perform each stage of restoration on it first. The Universals are considerably more numerous than Trans-Oceanics and are also less complex, both physically and electrically. Figure 10-1 shows a Zenith Universal Model 6G001, companion to the 7G605 Trans-Oceanics, and Figure 10-2 shows the Universal 6G001Y, companion to the 8G005Y Trans-Oceanics. Of the two Trans-Oceanics, the earlier 8G005Y is probably the best "starter" for a beginning restorer, even though this model is rather complex electronically. This radio was covered with the exact same Black Stag cloth that cloaked all of the black tube-type Trans-Oceanics.

Neither physical nor electronic restoration of Trans-Oceanics can be done successfully if you are upset, harried or in a rush. It cannot be done successfully without applying relaxed, unhurried concentration, a positive attitude and a very real appreciation for the artistry of the original designers and fabricators of the Trans-Oceanics. However, with care, caution, and above all, with the right attitude, practically anyone can restore a Trans-Oceanic to its full glory. You may, just possibly, become a better person at the same time. There really is Zen in the restoration of the Zenith Trans-Oceanics!

PHYSICAL RESTORATION

There are two broad steps in the physical restoration of tube-type Trans-Oceanics: 1) cleaning and minor repair, and 2) major or complete restoration. These remarks are organized first as a discussion of both steps which is applicable to most tube-type Trans-Oceanics. The general discussion is followed by specific notes and recommendations related to each Trans-Oceanic model. Study the general remarks first and then refer to the appropriate model-specific section. A list of the tools and materials helpful to the Trans-Oceanic restorer is found in Appendix I; suppliers for the specific products mentioned in the following discussion are presented in Appendix II.

CLEANING AND MINOR REPAIR

Other than a wipedown with a cloth dampened with Armor-All™, no cleaning ought to proceed until the chassis is removed from the cabinet. Removal of tube-type Trans-Oceanic chassis is virtually identical in all models:

1) Remove the tuning and volume control knobs. These are friction-fit knobs in all models and are removed by pulling straight out.
2) Remove the Waverod whip antenna. It is held to the side of the cabinet by two nuts which run on threaded studs. Draw an exact sketch of where the Waverod lead plugs into the chassis.
3) Disconnect the metal upper chassis brace from the inner top of the cabinet. Look for the brace directly above the rather fragile RF section on the left side of the chassis. The 1/4" wide strap braces the top of the RF tower and is supposed to be attached to the inner top of the cabinet by a Phillips-head screw (early models used a screw-attached wood block).
4) Remove the two hex-head screws which hold the chassis down on its shelf. The screw heads may be seen from the battery compartment below the chassis. The easiest way to remove these is by using a 1/4" nut driver inserted through the two holes in the bottom of the cabinet. Please do not use pliers through the battery compartment or you may destroy the screw heads.

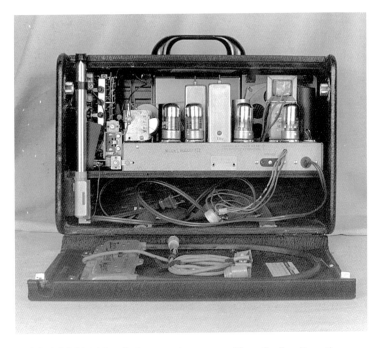

Model 8G005Y with the rear door open. Note the fragile coil tower behind the Waverod on the left and the broadcast band Wavemagnet in its storage position on the inner face of the rear door.

5) Slide the chassis *GINGERLY* out of the cabinet. Be *VERY* careful of the delicate RF section on the left with its small unprotected wires. Watch carefully so as not to catch any delicate parts on the two door latch mechanisms on the upper left and right of the cabinet.

As soon as the chassis is removed from the cabinet, turn it gently on its back and look for signs of excessive heat, areas of corrosion, melted wax, *etc.* Take notes on an index card which will stay in the battery compartment of the radio long-term, like a patient's chart in the hospital. These tell-tale signs will help you diagnose any electronic problems when that stage of restoration begins.

After the inspection and note-taking, you may proceed to clean the inside of the cabinet using cloth dampened with Armor-All™ or furniture wax. Take this opportunity to reglue all paper stickers, *etc.*, to the inner cabinet sides with white liquid glue.

The next step is to clean the exterior of the case. According to its condition, this usually entails a gentle scrub with soap and water, followed by drying and a liberal coat of Armor-All™.

The final step should be a thorough cleaning of the chassis top and sides and a liberal application of contact cleaner/enhancer to the controls. The best method to clean the chassis is to remove all of the tubes (*IMPORTANT:* if there is not a tube placement chart on the side-wall of the receiver cabinet, make one on an index card as you remove the tubes. The tubes are numbered on the glass side or top). After the tubes are removed, a soft cloth can be used to get the chassis top in "showroom condition." Then apply a contact cleaner/enhancer to all important electrical contacts. These would include the Radiorgan switches and the volume control. Operation will be improved by scraping each pin of each of the tubes to remove corrosion. This is followed by spraying the tube sockets with contact enhancer, as well. When dusting and cleaning the tubes, be careful not to remove the designation markings on the glass envelopes. The tubes are then reseated and worked back and forth a bit in the sockets to insure maximum contact.

You are now ready to reassemble the receiver or to proceed with a full physical or electrical restoration.

Disassembly

The first step in any serious work on the Trans-Oceanic cabinet is to at least partially disassemble it. This may sound daunting, but it is relatively easy and quick. Disassembly should be done even if the only planned restoration is to the Black Stag cabinet covering.

After removing the Waverod and chassis, the next step is to remove the plastic front panel of the receiver. In the H-500 and 600 Series receivers, the front panel is simply secured by five small screws visible from the front of the radio. Remove these screws and the panel comes off easily. Removing the front panel from 8G005Y and G500 radios is much more difficult and should be avoided if possible. Refer to comments in the model-specific section which follows.

The next step is to remove the main carrying handle. This is secured by two screws through the cabinet top. The cabinet front clasp should also be removed. The fixed lower section of the clasp is secured by two screws which may be accessed through the battery compartment. The movable upper part of the clasp is also secured by two screws. In the early models, these two screws are exposed and may be easily removed. The H-500 and 600 Series clasp screws are covered by paper; their removal is discussed in the sections dealing with model-specific restoration problems.

All interior hardware should also be removed from the rear door and interior cabinet sides. Since this hardware is all secured by screws, no major problems should occur. Be sure to draw a sketch or take a Polaroid photo to help you reassemble the hardware. You *will not* remember, when the time comes. It is also a good idea to store the hardware in a small container for safekeeping.

At this point, the cabinet should be free of all hardware except the hinged front and rear doors. The rear door is hinged with permanently attached hinges and should not be detached.

Although it is not always necessary, it is rather easy to remove, repair and adjust the front doors of all models of tube-type Trans-Oceanics. In all models except the 600 Series, the cabinet top consists of two pieces of plywood. The upper (outer) piece is covered by the Black Stag and conceals the hinges and anchor points for the front hinges which are mounted on the inner structural cabinet top. The outer finished cabinet top may be removed from inner structural top by backing out four wood screws. These screws may be easily seen at the corners of the inner top, *looking from inside the cabinet.* Refer to the model-specific comments for notes on removing the front doors of 600 Series cabinets.

FULL PHYSICAL RESTORATION

There are five main areas of full physical restoration to be accomplished: 1) the wooden cabinet, 2) the Black Stag cabinet covering, 3) the plastic parts, 4) the various pieces of brass hardware, and 5) the Waverod, plus other bits and pieces of interior hardware.

Repairing the Wooden Cabinet

The authors have rarely encountered a Trans-Oceanic which has been mistreated so badly that it needs major structural repair. However, an "open joint" along one or two cabinet edges is sometimes found. Since the cabinet is basically a plywood box, repair is relatively simple. Yellow carpenter's glue should be flowed or injected into the open joint and then the joint closed by the use of weights or bar clamps. Be sure to sponge off excess glue before it sets.

Restoring the Black Stag Covering

Scuffs, cuts, tears and holes in the Black Stag covering are the single most common "restoration problem." Ironically, it is also the most difficult to solve. In truth, neither the authors nor other experienced Trans-Oceanic restorers have a total solution to this difficult problem. To date, the best solution is to use white glue to reattach any torn and loose fragments of Black Stag after having gently washed the cabinet with a damp sponge.

The black paper (not Black Stag) which covers the inside of both the rear and front doors can be a special problem. Often it is delaminated and raised from the plywood in a half-inch wide rigid bubble which runs around the edge of the door. This seemingly unsolvable problem is easily conquered. Simply apply warm water to the stiff raised area with a sponge until it becomes pliable. Then, use white paper glue to reattach the black paper to the wood. It may be necessary to press the paper in place several times before the glue begins to bond.

Now that everything is back in place and there are no "loose ends," it is time to bring the cabinet as near to original condition as possible. Restorers have found that black aniline shoe dye (not liquid shoe polish) works very well in permanently darkening exposed wood or various light abrasions in the Black Stag. The dye can be obtained in an applicator at most shoe repair centers and many supermarkets and should be applied to all discolored areas. As the dye begins to soak in and dry, it is good practice to blend it in to the surrounding Black Stag with your hand or a soft cloth. NOTE: If Armor All™ has been previously applied, it must be removed with a good cleaner, such as Murphy's Oil Soap, or the black shoe dye will not apply uniformly. The intent of this dyeing process is to get every bit of the Black Stag and black paper as black as possible but without heavy shiny buildup or unevenness in the darkness of color. After everything is relatively evenly black, the surface of the Black Stag can be restored to a startling degree by liberal application of black paste shoe polish. Two or three medium coats work very well. Although this is not a perfect solution, and it may even sound amateurish, it is the process that professional restorers at major museums have found most effective in similar circumstances. It is also the method which was suggested by the factory. The goal is not to "spit shine" the Black Stag to a mirror finish. Rather, it should glow—with the raised areas of the texture being somewhat shiny and with the depressed areas being jet black, but not shiny.

Restoring Plastic Parts

Returning the plastic parts to their original soft sumptuous glow will do more for the appearance of your Trans-Oceanic than you might imagine. The most visual improvement comes from restoring the Wavemagnet and front panel of the receiver to their original condition. Before you start on these two items, however, two "cautions":

FIRST CAUTION: About half of the Trans-Oceanics we have seen have stress cracks in some portion of the front panel. Examine your front panel carefully for these cracks. We know of no cure for these cracks, but you do not want to make them much worse by inadvertently overenthusiastic cleaning and polishing.

SECOND CAUTION: Do not risk removing the silk screened lettering on the inner side of the clear dial face of the H-500 and 600 Series receivers by washing the inner side of the panel. The best advice is to simply clean the inner side of the dial window with a *dry* cloth. Mild soap and water *probably* will not remove the lettering, but it has been known to do so on some brands of radios. *DO NOT RISK IT*!

With these cautions in mind, a gentle cleaning of the Wavemagnet and front panel should proceed. Most restorations of these parts simply require removing a number of small scratches from the plastic and then bringing the entire finish from its time-dulled state to full color and shine. For these two jobs, the best product currently available is Novus Plastic Polish #1 and #2. Novus #2 is a very mild liquid abrasive and polish which will remove most minor scratches. After you have brought the surface back to a relatively even lustre, Novus #1 polish is applied. Besides bringing the surface to a high shine, this special formulation is antifog, anti-static, and a dust repellent. If your Trans-Oceanic received only the minor bruises of time, Novus #1 and #2 will reward you with "like new" Trans-Oceanic plastic parts with less than an hour of applied elbow grease.

If the plastic parts of your radio have suffered greatly, do not give up! With Wavemagnets, at least, it is possible to perform near miracles. It is even possible to remove paint from Wavemagnets. Proceed cautiously and use abrading rather than chemical techniques. Even mild mineral spirits will etch and dull the plastic Wavemagnet badly. The *harshest* abrading technique that you should use is using 000 very fine steel wool. A less aggressive technique is using a soft cloth with the finest available automotive rubbing compound and plenty of elbow grease. The intent of this process is to get all foreign matter removed from the plastic, to smooth out all scratches and gouges and to bring the entire surface to the same general dull sheen. That accomplished, you can begin to apply several thick coats of Novus Polish #1. If possible, pour the polish on the horizontal surface until it literally puddles. Left overnight, the polish will slowly dry into what looks like the original plastic finish. This new surface may need renewing with a light wipedown on an annual basis.

Most tube-type Trans-Oceanic models have inlaid lacquer lettering in the plastic of the front panel or the Wavemagnet. Restoring or replacing this thick colored lacquer is simplicity itself with a product designed for this purpose: "Lacquer-Stik." After thoroughly cleaning the plastic, rub the end of these 4" x 1/2" sticks of solid lacquer briskly across the incised surface. The friction of the rubbing slightly liquifies the lacquer and it is deposited in the incised lettering. Any excess lacquer may be wiped off the surface with a soft cloth. The Lacquer-Stik should be rubbed vigorously on some scrap wood or paper before doing so on the plastic. This wastes lacquer, but by pre-softening the Lacquer-Stik, it eliminates the possibility of scratching the plastic face of your radio while applying the lacquer. This entire process may sound daunting, but it is about the easiest to accomplish of any of the restoration steps.

The last two plastic items to tackle are the two plastic knobs and the carrying handle. They may be restored "by hand" as discussed above. However, if you have a small, bench-mounted grinding wheel, simply replace the grindstone with a cloth buffing wheel. The knobs and the carrying handle polish quite nicely using a medium grit buffing rouge to cut any scratches away and a mild buffing rouge for polish and color. The handle and knobs will literally glow in no time!

Restoring Brass Hardware and Trim

The brass hardware and trim are quite similar on all tube model Trans-Oceanics. There is even a high degree of part interchangeability between models (please refer to Chapter 12 on Parts Commonality as needed).

Much antique brass found on vintage luggage and radios was protected by a coat of transparent lacquer to prevent tarnishing and, in many cases, to modify the natural color of brass. The color modification was generally done by adding a small amount of orange lacquer to the clear base. This produced a darker, richer "dark yellow" color. With the exception of the H-500, this color modification was NOT done to any Trans-Oceanic brass. This is probably because Commander McDonald and industrial designer R. D. Budlong were committed to the "bright gold" color, along with black, to reinforce the "Royalty of Radio" concept so predominant in Zenith advertising of the day.

All Trans-Oceanics except the H-500 used totally clear protective lacquer on their brass parts. This clear lacquer may have "yellowed" somewhat over time in response to ultraviolet light. Your first task is to decide whether to refinish/restore the brass hardware at all. If you decide to refinish/restore any of the brass, you should commit to doing it all. If you do not refinish it all, there will be a noticeable color difference between the refinished and original "yellowed" brass. Again, however, the lighter yellow of the refinished brass is actually the original color.

Before you begin to refinish the brass, two cautions:

CAUTION: Although the handle hinge hardware is solid brass, the remainder of the brass hardware is actually thickly brass-plated mild steel or pot metal.

CAUTION: A few brass parts have either narrow black lacquer-filled or broad enamel-filled lettering on them. Refer to the model-specific restoration comments for notes on restoration of these parts.

The first step in brass refinishing is to totally remove the old protective lacquer coating. This may be done by an overnight soak in white vinegar or a brief bath in lacquer thinner. After removing the coating, you may choose to buff out all small scratches gently and by hand with 0000 steel wool. If the scratches are at all deep, it is best to leave them since you may buff completely through the brass plating.

The next restoration step is a thorough final cleaning and polishing using a very good commercial brass polish. Wash the brass in mild soapy water and rinse and dry thoroughly. The final step is applying a new protective coating. Most professional restorers use the high gloss clear acrylic plastic available in spray cans at most hardware stores. Be sure that you are spraying in a relatively dust-free, lint-free environment; otherwise, the coating will have a marred finish, and you will be doing the whole process again.

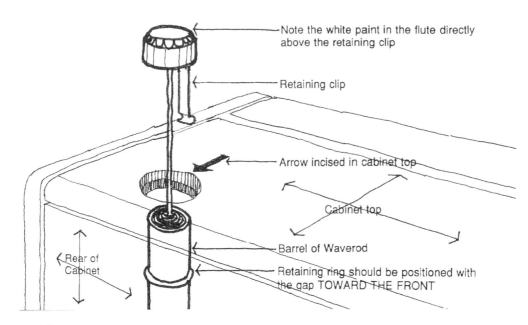

Note the white paint in the flute directly above the retaining clip

Retaining clip

Arrow incised in cabinet top

Cabinet top

Barrel of Waverod

Retaining ring should be positioned with the gap TOWARD THE FRONT

Rear of Cabinet

Figure 10-3

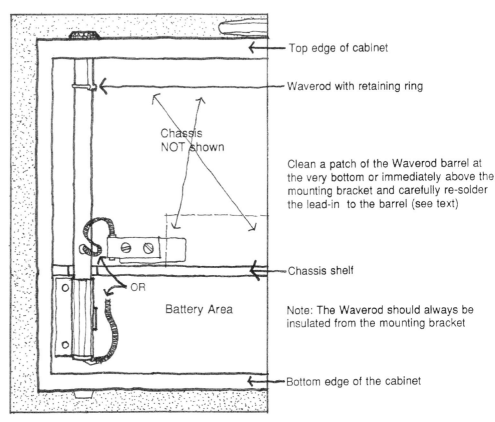

Top edge of cabinet

Waverod with retaining ring

Chassis NOT shown

Clean a patch of the Waverod barrel at the very bottom or immediately above the mounting bracket and carefully re-solder the lead-in to the barrel (see text)

Chassis shelf

Battery Area

OR

Note: The Waverod should always be insulated from the mounting bracket

Bottom edge of the cabinet

Figure 10-4

Waverod

The Waverod is certainly the single most vulnerable part of any Trans-Oceanic. About half of the Trans-Oceanics that we have seen have had seriously damaged Waverods. If you are lucky and have a Trans-Oceanic with an undamaged Waverod, you need only peruse the following two notes:

1) Commercial chrome cleaner/polish followed by lubrication with WD-40 works very well to rejuvenate otherwise undamaged Trans-Oceanic Waverods. NOTE: The Waverod of the 7G605 Clipper was nickel-plated but still responds to the chrome polish.
2) Remounting the Waverod takes special care which was often not given by electronic repair people, even in "the good old days." First, the Waverod was *ALWAYS* insulated from its mounting bracket. In the early post-war era, this was accomplished by a thin plastic sleeve the height of the mounting bracket. In later years, this sleeve was a section of clear polyethylene tubing with about 1/16" wall thickness. In any case, if this outer insulation between the Waverod and its bracket is missing, replace it with a continuous layer of plastic electrician's tape. The second item of special care is in the placement of the Waverod in its bracket (see Figure 10-3). First, examine the fluted plastic knob on top the Waverod. The flute that is directly above the thin metal retaining clip should have white lacquer in it. This is so that you can tell the radial (rotary) position of the retaining clip when the Waverod is completely retracted into the cabinet. Now note that there is a gap in the **retaining ring** on the upper barrel of the Waverod and note that the cabinet has a 3/4" long arrow incised into the top and pointing into the Waverod hole from the direction of the front of the cabinet. When you reattach the Waverod to the cabinet, make sure that the gap in the retaining ring faces the front of the radio (faces away from you as you reattach it). Now that the Waverod is properly positioned, you can turn the white dot of the Waverod knob directly to the front of the receiver (e.g., line the white dot up with the arrow on the cabinet) and the Waverod will predictably spring into action!

Waverod Repair

If the Waverod is bent, it is almost impossible to straighten. Even small bends in the solid (smallest) section of the antenna most often get worse rather than better when "straightening" is attempted. We recommend the acquisition of a replacement Waverod from a "parts radio," if at all possible.

There is one Waverod repair which is possible. The lead wire which carries the radio signals from the Waverod to the chassis is often broken off where it joins the Waverod. If the metal tab on the base of the Waverod is still present, simply strip the wire and resolder it to the tab. If the tab has been broken off, remove the Waverod from the cabinet and scrape a bare spot on a portion of the outer barrel of the antenna. Then pull all of the concentric cylinders of the antenna most of the way out of the outer tube (the barrel) for protection and solder the wire directly to the barrel. A small propane torch makes this job easier than using a soldering iron. Take care not to overheat the Waverod (see illustration, Figure 10-4).

If the damage to the Waverod is extensive but such that it can be collapsed and stored in its normal position, it will present a "normal" look from the outside of the cabinet. In such a case, the Trans-Oceanic can still be exhibited and even used on all bands with no one being the wiser. Broadcast band reception is handled by the Wavemagnet antenna, so it is never affected by the position or condition of the Waverod. Shortwave reception is possible without the Waverod by using a wire antenna. As little as 10 or 15 feet of insulated wire extended along a baseboard or shelf bottom will often suffice. If this proves unsatisfactory, a 20-to-50 foot outdoor wire antenna will usually work wonders. It is also possible to use an electrically amplified whip antenna (called an active antenna). These are found in many shortwave hobby catalogues. In any case, the input from these antennas goes to the screw connection in the rear of the radio chassis.

Bits and Pieces of Hardware

There are a number of bits and pieces of hardware associated with the interior of the Trans-Oceanic cabinet or the inner surface of the rear door. These include the two-part spring catches that hold the rear door

closed, the various mechanisms to store the Wavemagnet suction cups (usually two circular male snaps, either on a metal strip or individually screwed to the inner surface of the door), the various clips used to hold the Wavemagnet lead-in wire, various clips to hold owner's manuals and, sometimes, a 6" long "clip-strip" to hold a spare set of vacuum tubes. In various models these bits of hardware were "blued" like gun barrels, or painted or sometimes just left bare. The best restoration technique is to clean these pieces thoroughly, remove any rust and try to return them to their original finish. It is also good practice to then coat each of these parts with clear acrylic spray to inhibit rust. Careful attention to these details will really impact the appearance of your Trans-Oceanic when you reach the happy moment of reassembly!

PHYSICAL RESTORATION NOTES BY SPECIFIC MODEL

The following notes apply specifically to particular models of the tube-type Trans-Oceanics. Be sure to study the appropriate sections which apply to your Trans-Oceanic before beginning restoration.

7G605

Disassembly

Removal of the 7G605 chassis differs slightly from the other tube-type models because the Radiorgan is not attached to the chassis. To remove the chassis, take off the Radiorgan escutcheon plate and the knobs. The Radiorgan and its attached cable should then be pulled back so that it hangs from the back of the radio. The bottom chassis bolts are removed in the normal manner and an additional bolt securing the chassis from the right side of the cabinet (as you face the back of the radio) is removed. Remove the wooden retainer block in the vicinity of the coil tower and disconnect the Wavemagnet and speaker. The chassis is removed by retracting the right end enough to clear the cabinet and then applying side movement so that the chassis clears the Waverod support bolts.

Restoration

Proceed as described in the general restoration section. The imitation alligator hide is embossed and is more easily cleaned if a soft brush is used to apply the cleaning soap. The speaker grille cloth can be cleaned and restored with the careful application of a 50/50 dilution of commercial spot remover; check to be sure there will be no harm by testing the solution at the edge of the cloth before proceeding with general application.

Model 8G005Y

Disassembly

No special disassembly problems exist except the removal of the plastic front face. This can only be accomplished *after* removal of the chassis. The lower edge of the front panel is held in place with four screws accessible from the front of the radio (no problem here). The upper edge of the front panel is held in place by four rectangular spring clips forced down over plastic studs which are themselves a part of the front panel. It is difficult, but sometimes possible, to remove these clips using a sharp awl to bend their tines upward and outward away from the plastic stud. This may shatter the stud, however. If that happens, it is often possible to reinstall the front panel with tape or glue taking the place of the shattered stud. *PROCEED WITH GREAT CAUTION.*

Unfortunately, it is necessary to remove the front plastic panel before the brass dial escutcheon can be separated from the main front panel.

Restoration

Proceed as described in the general restoration discussion. The only really difficult task is cleaning the clear lacquer off the brass dial escutcheon *without* removing the black enamel from the word "TRANS-OCE-ANIC." It is possible to do this by using a soft cloth moistened with lacquer thinner and taking special care in the area of the letters. After the clear coating is removed, proceed as discussed previously to refinish the brass escutcheon, being careful to preserve the black enamel in the incised lettering.

Use white Lacquer-Stik to refill the lettering on the plastic front panel and black Lacquer-Stik to refill the script "Zenith" on the main latch.

G500

Disassembly

Refer to the previous 8G005Y notes for removal of the plastic front panel. Otherwise, there are no unusual problems. Note, however, that the pull-down "chin door" with hidden log book that existed on the 8G005Y was eliminated on this model, even though the cabinets appear identical.

Restoration

Follow the 8G005Y instructions for cleaning and restoration of the brass dial escutcheon. Note that the raised round Zenith corporate seal on the Wavemagnet touches the clear plastic dial cover when the front cabinet door is closed and latched. This usually has caused scratches in the middle of the plastic dial cover and also mars the Zenith corporate seal on the Wavemagnet. Novus Plastic Polish #1 is recommended for the dial cover.

Restore the round Zenith corporate seal as follows:

1) Carefully mask the circular seal so that all of the raised seal is exposed and the entire remainder of the Wavemagnet is covered.
2) Use commercial brass polish or other means (0000 steel wool, *etc.*) to clean and buff the corporate seal.
3) Wash the seal and recheck the masking.
4) Spray the entire seal with one or two coats of high gloss acrylic spray.
5) Unlike the 8G005Y, the rear face of the Wavemagnet was left unpainted, EXCEPT for the upper face of the letters in "Zenith" and both "Wavemagnet" words. These three words were silk screened gold lacquer.
6) Restore the remainder of the Wavemagnet as explained in the earlier general discussion.

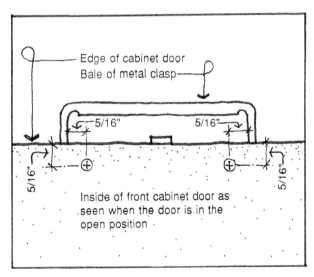

Edge of cabinet door
Bale of metal clasp
5/16"   5/16"
5/16"   5/16"
Inside of front cabinet door as seen when the door is in the open position

H500

Disassembly

There are no special problems in disassembling the H500 except for removing the upper portion of the cabinet clasp lock from the front cabinet door. Please refer to Figure 10-5 for exact locations of the two flat Phillips-head screws which hold the clasp to the lip of the front cabinet door. Their heads are obscured by the gray paper covering of the inner side of the door. Slit the paper with a *very sharp* (Exacto or scalpel) knife, carefully move the gray paper aside and remove the screws. When the H500 is reassembled, reglue the gray paper with white paper glue. If done properly, the two slits will be unnoticeable.

Restoration

IMPORTANT NOTE: The H500 is the one model where the brass hardware parts were coated with a **NON-CLEAR** lacquer. The color of all brass parts, including the brass knob inserts, is a much darker gold than the natural brass color used on other tube-type Trans-Oceanics. Most importantly of all, the gold paint used to "back paint" the central portion of the plastic front face of the radio *is the same darker gold color*. A restorer is faced with three options:

1) Do no brass restoration.
2) Restore the brass hardware "normally" as discussed elsewhere in this chapter *and have the brass be very noticeably lighter than the golden front face.*
3) Prepare a clear lacquer plus orange lacquer mix and spray the parts using an airbrush or other small paint sprayer.

The latter option, though considerably more trouble, will reward you with a restored Trans-Oceanic that looks *right* and is therefore highly recommended. Finally, refer to the brass restoration notes of the 8G005Y concerning removal of the clear coating without removing the word "Trans-Oceanic."

Restoring the inner surface of the front cabinet door and the Wavemagnet of the H500 also requires special attention. The covering of the inner side of the front door was a special gray paper selected by industrial designer Budlong. The color of this paper proved especially vulnerable to ultraviolet light; as a result, most H500's now have inner doors that are almost black. The original value (percent of black) of the paper was only slightly darker than the gray of the Wavemagnet. The color of the paper was somewhat variegated, as well, giving it a vaguely "leatherette" look. If a restorer wished to pursue the "like new" philosophy of restoration to its logical conclusion, the paper could be replaced by a near match. You may consult a specialty paper house; a gray "parchment" paper carried by many paper dealers is a relatively close match. Unless the original gray paper is *very* badly discolored or damaged, it is wiser to carefully clean the existing paper and leave well enough alone.

Restoring the H500's Wavemagnet is also a bit of a challenge. Plastic restoration can follow the process outlined in the previous general discussion. Restoring the H500's small all-brass corporate seal is a bit tricky, for it cannot be removed from the Wavemagnet. However, it is usually possible to slide pieces of stiff paper between the brass seal and the Wavemagnet. It is then possible to strip, clean and recoat the brass seal without damaging the Wavemagnet. Some of the enamel from the raised white lettering on the back side of the Wavemagnet is often found to have adhered to the gray paper that it rests on. At this point, we have not found a satisfactory way to restore this painted surface in a home-based restoration setting.

Finally, in "restored" H500s, the lacquer filling of the incised front side of the Wavemagnet is often done incorrectly. White lacquer fill was used for the words "Zenith" and "Wavemagnet." However, *gold* lacquer fill was used for the parallel vertical lines covering the lower 50% of the Wavemagnet. The original gold fill was apparently quite fragile, for it is often found totally missing. White filling of these vertical lines is seen from time to time; these instances are the work of misinformed restorers. Color advertising and photos exist which clearly show these vertical lines filled with *gold*.

600 Series

Disassembly

The only difficulty in disassembling the 600 Series receiver is removing the plastic door/log-book fittings from the inner surface of the front cabinet door. Follow this procedure:

1) Place the front cabinet door in an upright position, as if to use the radio.
2) Open the rectangular plastic log book drawer.
3) Find and remove six small flathead screws within the space revealed by the open log book door.
4) Slide the entire plastic assembly upwards, vertically, in the same plane as the front cabinet door. Movement of about 1/4" vertically releases the plastic from two hidden clips which assist the screws in holding the plastic fitting on the inner surface of the cabinet door.

The front cabinet door of the 600 Series is removed differently but easily from the main cabinet. After the chassis is removed, it is easy to spot oval-shaped straps sticking down into the radio compartment. Look at the upper portion of the inside of the cabinet side panels. These straps extend up through the cabinet top from the inside and connect to the door hinge itself. Two screws hold each strap to the cabinet. Remove them and the straps slip through the top of the case as the door is lifted off. Refer to the commentary and figure in the H500 notes above for the method used to remove the upper portion of the cabinet front door latch.

Restoration

There are no unusual aspects to the restoration of the Black Stag covered version of the 600 Series Trans-Oceanics. *Please note, however, that unlike the H500, the horizontal grooves in the plastic covering of the front cabinet drawer were NOT filled with lacquer.*

Please refer to the brass restoration notes under model 8G005Y for notes concerning removal of the clear lacquer coating on the lower portion of the main cabinet latch. It is necessary to remove the coating while preserving the black lacquer fill in the word "Trans-Oceanic."

600 Series Leather Version

A well-preserved or restored **leather** version of the Trans-Oceanic 600 Series simply has to be one of the most beautiful radios ever built! Restoration of any antique leather item including the covering of a radio is a highly specialized art. If you have a local saddlery of good repute, particularly one that has been in business for many years, we strongly recommend that you take your leather Trans-Oceanic to them and ask for their advice on restoration of the leather.

Disassembly

Please refer to the section on the 600 Series above for disassembly notes.

Restoration

The plastic parts of a leather Trans-Oceanic were all dark or light brown to match the cordovan brown leather rather than the standard black or gray. All of the plastic parts are *dark* brown except the movable log book door on the inner surface of the front lid, which is *light* brown. The background color of the slide rule dial face is also dark brown. The only difference between the two chassis designations, 6L40 (black) and 6L41 (brown) is the plastic color.

If you do not have access to good advice on leather restoration, here is a direction along which you may cautiously proceed:

1) After stripping the cabinet of all hardware except the two sets of door hinges and doors, gently wash the entire leather covering using saddle soap.
1A) Some restorers use numerous applications of the semi-solid mechanics hand cleaning paste cream to clean very dirty leather Trans-Oceanics. This is sold under several trade-names including "Go-Jo" and "Goop." *PROCEED CAUTIOUSLY AND EXPERIMENT ON THE UNDERSIDE OF THE CABINET FIRST.*
2) After the leather is thoroughly dry, re-cement any loose areas of the leather covering with contact cement. Re-cement the paper-covered areas (the inner surfaces of both doors) using white glue.
3) Apply a *light* coat of Neat's Foot Oil or Mink Oil to rejuvenate the leather. *Allow this to soak in for at least 48 hours before proceeding. NOTE*: Neat's Foot or Mink Oil will cause a slight darkening. It can also cause a spotty coloration unless the leather cabinet has been cleaned thoroughly.
4) There will probably be abraded areas of leather and long scratches in the leather which appear much lighter than the surrounding cordovan color. These light areas and lines can usually be brought to an approximate match with the main cordovan color by applying "NEUTRAL" paste shoe polish directly to the affected areas. *TEST THIS ON THE BOTTOM SURFACE OF THE CABINET FIRST AND ALLOW TO CURE AT LEAST OVERNIGHT.* Applying cordovan rather than neutral color paste wax to these lighter areas turns them permanently much darker than the surrounding cordovan color.
5) After you are satisfied with the general coloration of the leather surface, apply one or two coats of cordovan color paste shoe wax to the leather portions of the cabinet.

Careful restoration of the leather portions of your Trans-Oceanic will bring it to its full beauty and reward your care with many years of pleasure. Count yourself among the lucky few to own and care for one of the most beautiful products of the electronic age!

ELECTRONIC RESTORATION

A radio receiver is composed of a number of electrical and non-electrical components that function to take radio waves from the air to create a clear and faithful reproduction of the program that came from the transmitter. A number of these components may wear out or go bad with age and use, and may need replacing to render a "dead" radio once again a valuable and useful item. A Trans-Oceanic is just like an automobile in that it will not function properly without fully functioning parts. With caution, some electronic repair and restoration of tube-type Trans-Oceanics may be done by non-technical radio owners. It is not the intent of this section to turn the reader into a certified electronics technician. The following sections present several levels of discussion of the most common electronic problems encountered in three-way portables such as the tube-type Trans-Oceanic and provide some insight into their repair and restoration. This format is intended to be of assistance to all readers, no matter what their level of technical expertise.

Safety First

Unlike now-familiar solid-state devices, all tube electronic gear is often extremely hazardous to operate and, especially, to repair. These hazards fall in two categories: fire hazard and life-threatening electrical shock hazard. No matter how rudimentary or advanced your electronic knowledge, please read, consider, and heed the following paragraphs.

Fire Hazard

All electrical devices are inherently low level fire hazards. However, three-way (or AC/DC) radios like the Trans-Oceanic are especially hazardous for several reasons. First and most importantly, a tube-type Trans-Oceanic is *not* really turned off when you turn it "off." For some very valid reasons, Trans-Oceanics were designed with the power switch "downstream" beyond the first few critical components. In other words, even when the power switch is off, the rectifier (tube or selenium type) and the electrolytic filter capacitors are under *full* voltage as long as the plug is in the wall socket. Selenium rectifiers, the first solid-state rectifiers, were used on all Trans-Oceanics using miniature tubes (G500, H500 and 600 Series). These selenium rectifiers were relatively failure-prone even when new. Would you bet your house, apartment or life on a 50-year-old selenium rectifier? No tube-type Trans-Oceanic should EVER be left plugged in when not in use unless a number of simple modifications have been made. These modifications will be discussed later.

A second fire hazard exists because, with the exception of the military model, R-520/URR, none of the tube-type Trans-Oceanics have a safety fuse. This lack of a fuse was standard practice in the 1950s and Zenith should not be faulted. However, operating 50-year-old electronic devices today without some fuse mechanism is very foolhardy. Again, solutions for this problem are discussed elsewhere in this chapter.

Shock Hazard

Most modern solid-state devices operate on 12v DC or less. Some tube-type radios have internal voltages as high as a lethal 600 volts! We are fortunate that no voltages in tube-type Trans-Oceanics exceed 117 volts. Be aware, however, that it is possible for 117 volts to be lethal. Further, due to the design of three-way radios, there is a good chance (sometimes as high as 50-50) that 117 volts of electricity will be present on the interior *metal* parts of the Trans-Oceanic *even when it is turned off.* To reiterate, if the set is plugged into the wall outlet, **it is possible to receive a lethal shock simply by opening the back of the cabinet and touching one of the bare metal parts!** There are only two ways to insure that the chassis is not charged with 117 volts: 1) unplug the set for a few minutes, or, 2) use an "isolation transformer" which breaks the potential direct electrical connection of the chassis to the power line.

Unlike most transformers which step the voltage up or down as current passes through them, an isolation transformer does not. If 117v AC is applied to one side of the transformer, 117v AC comes out the other side, yet the two sides are "isolated" from each other. These devices are available at almost any electronic parts house. An isolation transformer rated at 117v AC and .4 amps or higher will power a tube-type Trans-Oceanic. It would be desirable and convenient to wire in a switch and install a 3/16 or 1/4 amp slow-blow fuse in conjunction with the isolation transformer. This simple bench device can save lives. A Trans-Oceanic (or any three-way radio) should not be serviced without using an isolation transformer.

Finally, since 117 volts is present on a few parts of the circuitry and 70-90 volts is present through much of the radio, good safe and cautious practice is absolutely necessary at all times when working on these tube-type radios. Accidents can and do happen and none of us is either perfect or immune.

How Does a Tube-Type Trans-Oceanic Operate?

Tubes are the heart of the tube-type Trans-Oceanics. A tube, or more properly, a **vacuum tube**, may perform a variety of functions depending on its internal composition. A rectifier tube in the power supply, for example, changes alternating current (AC) from the wall outlet to the direct current (DC) needed for radio operation; several amplifier tubes increase the intensity of a desired signal at various points in the circuits.

Tubes are found in a variety of shapes and sizes. The 7G605 and 8G005Y Trans-Oceanics employ fairly large glass tubes with loktal bases (so named because the center pin of the tube snaps into a locking device in the tube socket which is designed to hold the tube firmly in place). The G500, H500 and the 600 Series models use miniature tubes.

The tubes of the Trans-Oceanic function in a specialized circuit called the **superheterodyne** circuit, designed by Edwin H. Armstrong in 1918. This circuit basically converts the modulated signals entering the radio to a common lower frequency which is more easily amplified or otherwise processed by the radio circuitry. As a modulated radio signal encounters the antenna of a Trans-Oceanic, it is transformed to a weak alternating current signal with the same frequency as the modulated radio wave. This weak signal is amplified by the radio frequency (RF) amplifier (tube type 1LN5 in the 7G605 and 8G005Y, and 1U4 in the newer Trans-Oceanics). The converter tube (ILA6, ILC6 or IL6) changes the signal from the RF amplifier to a fixed lower frequency called the intermediate frequency (IF) which is easier for the circuit to work with than the normally higher radio frequencies. The IF frequency for most radios (and all tube-type Trans-Oceanics) is 455 kHz. The IF carrier signal, still containing the audio signal from the transmitting station, is greatly amplified by the IF amplifier tube (1LN5 or 1U4) and sent to the detector tube (1LD5, 1S5 or 1U5), which separates the audio signal from the unwanted IF carrier.

The receiver at this point has the "same" audio signal that originated at the transmitter, but it is not strong enough to drive the speaker. An audio output (amplifier) tube (3Q5, 1LB4 or 3V4) is employed to amplify the signal to drive a speaker. The entire radio is powered by the power supply section, which receives AC from the wall outlet, rectifies (converts) it to produce DC and all the necessary voltages to operate the system. An additional tube was added to the Trans-Oceanic beginning with the 600 Series, the current regulator. The 50A1 regulator maintains a constant current to the filaments of the other tubes even when line voltage fluctuates as low as 90 volts or as high as 130 volts. Although this description of receiver operation is abbreviated and does not speak to a variety of other functions being performed in the circuitry, it should provide enough information to help the reader appreciate the "magic" that is going on inside the case of his or her Trans-Oceanic.

Help for Electronic Beginners

It is beyond the scope of this book to delve further into basic electronic component and circuit theory; however, all is far from lost. There are several current books in print written to introduce electronic neophytes to the not so mysterious world of electronic radio repair and restoration. These include *Old Time Radios, Restoration and Repair* by Carr and *Fixing Up Nice Old Radios* by Romney. Both of these books are excellent learning tools to guide you along the path to electronic literacy.

One other book, regrettably long out-of-print, is especially helpful: *Practical Radio Servicing* by Marcus and Levy. This book was written for what we would now call vocational-technical students. It was a basic text used in the 1940s and 1950s to help these students become professional radio repairmen. Besides being written at a very comfortable and practical level, *Practical Radio Servicing* contains an entire chapter on the repair of "three-way" radios such as the Trans-Oceanic. Although not in print for many years, this book can be easily obtained at a public library. If your library does not have it, they can obtain it through Inter-Library Loan. Another excellent book to look for in your library is *Elements of Radio Servicing* by Marcus and Levy, a more advanced edition of *Practical Radio Servicing* (refer to the Bibliography section at the end of this chapter for complete bibliographic information).

First Steps

In any radio electronic restoration, the first task is simply to get the radio operating, no matter how poorly, before proceeding with a restoration intended to return it to long-term usefulness. This two-step process, repair followed by in-depth restoration, is actually much easier and simpler than combining the two processes. Anyone with a modest amount of electrical understanding and a few tools can perform many of the early steps of both repair and restoration. The latter phases of both processes usually require more understanding, but even beginners will gain useful insights from the remainder of this chapter. Before proceeding, however, please reread and heed the section on safety.

STEP 1. When thinking of buying a tube-type radio, you should always try to find out as much about its electrical condition as possible.

STEP 2. Whatever its condition, the best practice is to run several checks first, before ever powering up your new Trans-Oceanic. You should remove the chassis from the cabinet (discussed previously) and do a complete visual check using this brief check list:

A) Power cord. Check for brittleness and disintegration of the plastic insulation, particularly near the plug and near the chassis. If there is any problem at all with the power cord, it must be replaced first.

B) Check the top of the of the chassis for any abnormalities. If there are tubes missing, they must be replaced. Corrosion near the (usually) black cardboard cylinder which contains the electrolytic capacitors is a bad sign.

C) Check the top of the chassis shelf in the wooden cabinet for any melted wax or corrosive residue. Either or both of these are also bad signs and may indicate major repair work before applying power to the chassis.

D) Check the underside of the chassis for obviously burned or melted parts. Again, anything appearing "odd" may indicate necessary repairs before proceeding.

STEP 3. If everything appears normal, you may then begin to apply power to the chassis. The best practice--one that all professional restorers use--is to apply power to the chassis *VERY SLOWLY*. This is usually done with a "variac," a variable voltage AC transformer. [A poor second choice tool for this process may be constructed by inserting a 1000 ohm 15-20 watt variable power resistor in one side of the AC power line between the wall outlet and the radio.] Voltage should be started around 50 to 70 volts and be brought up to 117 volts slowly over at least four hours.

The primary reason for this strategy is to "reform" the electrolytic capacitors in the set. These important capacitors tend to lose their effectiveness if they sit unused for a long time. The slow application of voltage will often return them to service. Another reason for the slow application of voltage is to be able to shut things down quickly if a component begins to overheat. *Obviously, this process should never be left unattended.*

If you have neither a variac nor a variable power resistor adequate to "bring the set up slowly," it is acceptable just to plug it in. It would be very smart, however, to have done all of the other First Steps before doing so. It is also very smart to do so with the chassis out of the cabinet and lying on its side so that you see both the bottom and top simultaneously.

Three Things to Remember

1) With 1.5 volt vacuum tubes, you *do not* see a glow from the filaments when they are working.
2) Connect the speaker (in the cabinet) to the appropriate wires with jumpers or you will not hear any sound even if the set is working properly.
3) Watch carefully for small spindles of smoke.

As the voltage reaches about 85-90 volts, you may be rewarded with sounds coming from an operating radio. If so, you can begin to think about partial or full electronic restoration. If not, it is time to deal with the most common causes of problems in tube-type Trans-Oceanics.

REPAIRING COMMON PROBLEMS

Tubes

The filaments (or "heaters") of the tubes in the Trans-Oceanic are connected together much the same as the old fashioned Christmas lights (in series) so that "if one goes out, they all go out." Filaments do indeed burn out occasionally and can cause receiver malfunction. It is more likely, however, that the malfunction of other elements in a single tube is the cause of your new Trans-Oceanic not working. Tubes may lose their vacuum, internal elements may cease functioning or lose their efficiency, or the failure of another component may cause overheating or shorting of the tube. Weak or nonfunctioning tubes must be replaced to restore radio operation. To insure that the new tube is not destroyed by excessive voltages allowed by failure of other components, it is necessary to investigate the chassis carefully for signs of component overheating or burning as discussed above.

If you suspect that a tube is not functioning properly and you are satisfied that there are no overheated or burned components, you have two immediate options: test or replace. If a tube tester is not available to you, and you wish to use your radio, it would be best to replace all tubes. The most frequent cause of a nonworking Trans-Oceanic is a bad tube. Many restorers purchase a single set of new tubes that they use in the radio they are demonstrating or using, and test and replace tubes in their other sets as needed. The ultimate, of course, would be to find a full set of Zenith brand tubes. Vacuum tubes are available from a number of commercial sources as well as at ham fests (a sort of flea market of radio "stuff"--also a good place to uncover a "new" Trans-Oceanic).

Common Radio Troubles Caused by Bad Tubes

(Tube numbers relate to the miniature tube chassis but may be transposed to other chassis)

Observed Effect	Probable Cause	Remedy
"Motorboats" on broadcast band	1U4 IF tube defective	Replace tube
	1U4 RF tube defective	Replace tube
Microphonic distortion on BC band	1U5 det./amp defective	Replace tube
Distorted audio on broadcast band	1U5 det./amp defective	Replace tube

Obtaining Service Information

Other than replacing the power cord and defective tubes, there is little electronic repair or restoration that anyone can do without a schematic circuit diagram and parts list. All tube-type Trans-Oceanics came with an Operators Manual and a large format (usually about 11" x 24") schematic circuit diagram and service sheet. If you got this sheet with your Trans-Oceanic, you are most fortunate. If not, you need to find at least a photocopy of one. Possible sources include fellow hobbyists and local hobby clubs, large libraries and some electronic supply houses in large cities.

In the 1950s and 1960s, these diagrams were also available from a commercial service which catered to radio repair shops. Howard W. Sams Co. of Indianapolis is still in business but no longer stocks these tube-type schematics. Sams "PhotoFact" service sheets are, however, more readily available than the original ones from Zenith. These books of service sheets are found in many large libraries. They are also often available as single sheets from a variety of hobby sources. Refer to Figure 10-6 for a list of tube-type Trans-Oceanic service data.

Capacitors

After vacuum tubes, the next most frequent cause of a nonworking old radio is failure of one or more capacitors--and a typical Trans-Oceanic may have 12-20 capacitors. A capacitor is a device that stores or releases electrons as they are needed in a circuit, much the same as a water tower stores and releases water. When a capacitor malfunctions in

MODEL NUMBER	CHASSIS NUMBER	YEAR PRINTED	ZENITH SERVICE MANUAL VOL-PAGE	SAMS PHOTO-FACT SET - FOLDER
7G605	7BO4	1942	2-311	#190*
8G005	8C40	1946-48	3-96	467-33
8G005YTZ1	8C40TZ1	1946-48	3-100	53-27
8G005YTZ2	8C40TZ2	1946-48	3-100	53-27
G500	5G40	1950-51	4-393	83-16
H500	5H40	1951-53	4-423	152-12
L600Y	6L40	1954	5-31	254-13
L600L	6L41	1954	5-31	254-13
R600Y	6R40	1954	5-96	254-13
R600L	6R41	1954	5-96	254-13
T600Y	6T40	1955	5-122	254-13
T600L	6T41	1955	5-122	254-13
Y600Y	6T40Z	1956	6-201	254-13
Y600L	6T41Z	1956	6-201	254-13
A600Y	6A40	1957	7-73	381-16
A600L	6A41	1957	7-73	381-16
B600Y	6A40	1958	7-73	381-16
B600L	6A41	1958	7-73	381-16

*Not Sams PhotoFact; Published in *Manual of 1942 Most Popular Service Diagrams* by Supreme Publications.

Figure 10-6. TUBE-TYPE TRANS-OCEANIC SERVICE DATA

an electrical circuit, the required amount of electrical energy will not be available to operate the radio properly.

Capacitors may be constructed of a variety of materials, but no matter how they are constructed, the electrical effect is the same. In its simplest form, a capacitor is composed of two metal plates separated by an air insulator. The electrical size of the capacitor is termed its **capacity**; the larger the plates, the higher the capacity. Capacity is expressed in units called **farads**. As AC or pulsating DC passes through the capacitor, electrons build up on one plate, leaving a shortage on the other. As the low point of the AC cycle comes to the capacitor, the situation is reversed and electrons pile up on the other plate. In this fashion, the capacitor becomes charged and stores electrical energy. The capacitor thus stores or blocks the flow of DC energy and passes AC energy.

The most common capacitor failures in Trans-Oceanics occur in the electrolytic type capacitors that we discussed "reforming" earlier. Electrolytics, as they are usually called, are found in the power supply of your radio where they function to remove the ripples from the pulsating DC that comes from the rectifier (remember, the purpose of the rectifier is to convert bi-directional AC to the single direction DC needed to operate the radio). If the pulsating ripples which remain after rectification are not removed, they cause an audible hum. The electrolytics of the power supply are called filter capacitors and function to keep this hum from destroying radio reception.

The electrolytic capacitor differs in construction from the other most common type, paper capacitors, in that a paste of electrolytes is spread between the metal plates, providing a much higher capacity and thus higher value capacitors in smaller size. The paste dries out with storage and electrolytics may not function properly until current has flowed through them for a period of time--and often, once dried out, will never function properly again. The effect of the electrolyte gives the capacitor a positive and a negative end and this polarity **must** be observed if a replacement capacitor is installed--if installed backwards, the capacitor may explode. Although electrolytics may be the size and shape of paper capacitors, filter capacitors are usually found in "cans," aluminum or cardboard housings that contain several capacitors. In most tube-type Trans-

Oceanics, the filter capacitors are found on the right side of the upper chassis as you look from the back, and the can is covered by black cardboard.

When replacing capacitors, remember that the same or higher working voltage must always be used; for filter capacitors, however, the capacitance values may be two-to-three times the original value. It is always best to use the exact value for a replacement, since the original circuit was designed to contain that value.

Modern electrolytics are much smaller than those that were used originally in the Trans-Oceanic. If you find that any of the electrolytic capacitor sections are bad, all electrolytics should be replaced. To preserve the appearance of the chassis from the top side, and the value of the radio, *please* leave the existing electrolytics in place and secure the new small electrolytics to the underside of the chassis. There is plenty of room.

If you know that the tubes are good and the electrolytics seem OK, and your new Trans-Oceanic does not yet work even minimally, the fault is most likely one or more bad paper capacitors. Because they are inexpensive to construct and are not made of long lasting materials, they often break down after a few years of use. The most frequent breakdown is failure of the paper insulator, which then allows electrons to pass between the foil plates, creating an electrical **short**, which defeats the storage function of the capacitor. If the capacitor does not have a total or "dead" short but still allows some flow of electrons between the two plates, it is said to be **leaky**. Although a leaky capacitor is not as bad as a shorted one, it is still not functioning as it was designed to function. The third way a paper capacitor may fail is breakage of the leads where they attach to the foil plates. This creates an **open** capacitor and in the electrical circuit behaves the same as a cut wire. If a paper capacitor shorts or has severe leakage problems, it will often overheat, melting the waxy covering and leaving a telltale sign of a problem. This meltage may also occur if some other component has failed and allowed excessive voltage to pass through the capacitor. Refer to the section on Total Electronic Restoration for further comments on replacing paper capacitors.

Since we are now just discussing getting the Trans-Oceanic to work minimally, all that is necessary is to track down the faulty capacitors. This is relatively easy to do if you have a few tools and a schematic diagram. The diagnostic routine, however, is beyond the scope of this book. The books mentioned earlier in this chapter are especially helpful in diagnosing problems with capacitors. Paper capacitors are no longer used in consumer electronics (thankfully). For replacement capacitor types, please refer to the restoration discussion which follows.

Rectifiers

The mechanism in the Trans-Oceanic power supply which converts alternating current (AC) to direct current (DC) is called a rectifier. Until the mid-1950s, almost all AC-powered tube radios used a rectifier tube to do this conversion. The rectifier tube on the 7G605 and 8G005Y was the 117Z6G tube. The 8G005YTZ1 (and TZ2) used a modern miniature tube, the 117Z3. The G500 model ushered in the all-miniature tube chassis AND the first solid-state rectifier used on a Trans-Oceanic, the selenium rectifier. This is not a tube but a solid-state semi-conductor rectifier. Look for it as a sandwich of alternating big and small square metal plates. It is an approximately one inch cube and is screwed to the underside of the chassis. By modern standards, these selenium rectifiers are particularly prone to failure. Since they are one of the components that is under full voltage even when the Trans-Oceanic is left turned off, they are a fire hazard and should be replaced with a modern silicon diode (very small and inexpensive) such as the 1 amp 600 volt unit number IN4005. Be sure to install the new diode with the correct polarity as shown on the circuit schematic. Again, to preserve the historic value of the radio, it is an excellent idea to leave the large selenium rectifier physically in place. Just remove it from the electrical circuit and replace it, electrically, with the small silicon diode.

Other Parts Which May Fail

There are only two other major types of electronic components in the tube-type Trans-Oceanic: resistors and coils. Luckily, these two component types are not major sources of problems. However, both resistors and coils do fail from time to time.

Resistors

Resistors are the usually small tubular objects with the colored bands under the chassis. Resistors function by providing an electrical path that has a definite amount of opposition to current flow, the amount of resistance indicated by the color coding of the bands. In a practical sense, a resistor drops the voltage in a circuit. All the small color-banded resistors (known as **carbon resistors**) in the tube-type Trans-Oceanics are rated at 1/2 watt and are readily available. Carbon resistors tend to "last a lifetime." If they do fail, however, the result is usually a break in the current flow, much the same as if a wire is cut. Resistors fail in three ways: they may break, burn out, or change their resistance values.

Wire-wound resistors are found in the power supply circuits where they handle heavy currents. As the name implies, these resistors are constructed by winding resistance wire on a ceramic, glass or fiber core. The main resistor of this type in the Trans-Oceanic is the filament dropping resistor which drops the 117-volt line voltage to the 7.5-volt voltage needed by the tube filament circuit. This large resistor may burn out, but this is an infrequent occurrence; if it opens, the tube filaments will not light and the radio will not work. Other high wattage wire-wounds and carbon resistors may be found in the various models of the Trans-Oceanic.

Again, diagnosing specific problems associated with resistor failure is beyond the scope of this discussion. Any one of the books discussed earlier in this chapter will help in tracking down the occasional resistor-related problem.

Coils

A number of **coils** are found in the Trans-Oceanics. A coil is basically a length of wire wound around a core of some sort. Coils oppose any change in the current flowing through them by producing in them-

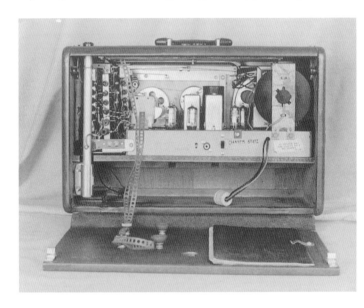

600 Series model Trans-Oceanic with the rear door open. Note the fragile coil tower behind the Waverod on the left and the power cord take up reel on the right.

selves an AC voltage, creating what is termed **inductance** The inductance of a coil may be changed by manipulating the type or position of the core or by changing the number of turns of wire or the shape of the coil. In a very simplistic sense, the coils of the Trans-Oceanic determine the operating frequencies of the receiver.

The most dominant coils of the Trans-Oceanic are found in the "coil cluster" at the extreme left of the chassis as you look from the back. Three sets of individual coils for each band are found in this cluster and allow the Trans-Oceanic to be an "all-band" receiver. The cluster is somewhat delicate and extreme care should be taken to protect these coils when you are removing or replacing the chassis. The most obvious indication that one of these coils may need attention is that the receiver does not function on one of the bands but functions on the others. Note also, though, that this same fault may be caused by a dirty bandswitch. This cluster of coils and a few nearby variable trimmer capacitors are adjusted during "RF alignment" discussed elsewhere in this chapter.

If problems exist within individual coils of the cluster, they are usually limited to breakage of the very fragile lead wires. These may often be reconnected if you can perform delicate soldering. Please note that there are no new replacement coils available. If one or more of these coils is irreparably damaged, it must be replaced with a coil from an appropriate "parts radio".

All Trans-Oceanics, indeed all superheterodyne-type radios, have one other set of adjustable coils: the **IF transformers** ("transformer" as used here means simply sets of specialized coils). These IF transformers tune exactly as the other coils except they function to adjust the intermediate frequency (IF) to precisely 455 kHz. These should not be disturbed unless you know what you are doing. If you feel that some adjustment may be needed, be certain to keep track of how many turns or fraction of turns you make so that you may return them to their original setting.

By This Time....

By this time, you should have found and replaced the defective tubes, capacitors, rectifiers, resistors and coils in your new Trans-Oceanic. It should be playing merrily away at least on the medium wave (AM) broadcast band. If you have a Trans-Oceanic which operates well on the AM band but not at all on any of the shortwave bands, your set has fallen prey to the most common fault found in tube-type Trans-Oceanics: dirty contacts on the AM band portion (YES, THE AM BANDSWITCH!) of the bandswitch tower. For rather esoteric engineering design reasons, all *shortwave* signals pass through a second set of contacts on the AM bandswitch before going on to the contacts of a particular shortwave band. This second set of contacts on the AM portion of the bandswitch is only closed when the AM band is *not* in use. Most Trans-Oceanics were actually used primarily on the AM band; when they were put in storage, most Trans-Oceanics were left with the AM band contacts engaged. This left that *second set* of contacts on the AM bandswitch (through which all shortwave signals must pass) in an "open" condition, completely exposed to long-term corrosion. The contacts are likely so corroded that they will not conduct the relatively weak radio signal on down the circuit path.

The solution to this problem is to spray the entire bandswitch tower (especially the AM bandswitch) with the best contact cleaner-enhancer available and to vigorously "cycle" or "exercise" the bandswitch by pushing button after button. Do not give up on this technique. It may take ten minutes, but the contact will almost always begin to work.

With any luck at all, you now have a Trans-Oceanic that works reasonably well and we are ready to contemplate the last two major activities: electronic restoration and alignment.

TOTAL ELECTRONIC RESTORATION

Once your new Trans-Oceanic has been minimally repaired and plays on all bands, you must consider just how much electronic restoration to do or have done. Although terminology and definitions vary a good bit in this area, there are really only two levels of electronic restoration: partial and complete. In the case of tube-type Trans-Oceanics, a partial restoration would include replacing all parts which affect the safety of the radio: the rectifier and all electrolytic capacitors. A "full" electronic restoration would include changing all of the above plus replacing all paper capacitors.

As discussed in the section "Safety First," the circuit design of tube-type Trans-Oceanics is such that the electrolytic filter capacitors and the rectifier (tube or selenium) are under full voltage as long as the set is plugged into the wall outlet. By modern safety standards, this is not terribly good practice, even with new components; having 50-year-old capacitors and rectifiers under full voltage at all times is an invitation to disaster.

Replacing the electrolytic capacitors with safe new units was discussed earlier in this chapter and is the very minimum electronic restoration that should be done. If your Trans-Oceanic has one of the failure prone selenium rectifiers, it should also be replaced. Please perform these relatively simple and quick substitutions before beginning to use your Trans-Oceanic on a continuing basis.

As stated earlier, the 10 to 20 paper capacitors in most tube-type radios are now "failures waiting to happen." If you choose not to replace all the paper capacitors, you, or someone, will have to troubleshoot a nonworking Trans-Oceanic often to determine which capacitors have failed. Further, failure of some of these inexpensive components can cause other components to be irreparably damaged at the same time.

A variety of new types of capacitors are available to replace the old paper ones. Electrically, any capacitor of the proper or near proper capacitance value and the same or higher working voltage is acceptable as a replacement. These may include ceramic, silver mica, or disc ceramic capacitors. A popular replacement capacitor is the "orange drop," a small, very efficient universally available capacitor. Aesthetically, however, the capacitor choice is not as broad. In order to maintain the original "look"

of the radio, like-appearing capacitors need to be installed, but the old style paper capacitors are no longer available. A few suppliers sell ceramic or metalized film tubular capacitors which are about the same size and shape (but not the same color) as the paper capacitors and make an acceptable substitute.

When replacing capacitors, remember that the same or higher working voltage must always be used. Most capacitors available at shopping center stores such as Radio Shack are generally low voltage replacements for modern electronics and will not work in your tube-type Trans-Oceanic. As for capacitance values, values within 20% are acceptable. For example, a .022 mfd capacitor may be used to replace a .02 mfd capacitor. Again, it is always best to use the exact value for a replacement, since the original circuit was designed to employ that value.

It is not necessary to replace any of the rectangular brown mica capacitors, if your set has any; these capacitors almost never fail. Likewise, complete replacement of resistors is never done. Although these components do change value radically or fail occasionally, the failure rate is so very low that complete replacement is simply not worth the time.

A last caution: When buying radios from fellow radio enthusiasts, inquire specifically about just how much electronic repair and restoration has been done. Even if you have been told that the receiver has been "totally re-capped," inquire specifically as to whether the paper capacitors have been replaced. "Totally re-capped" to many less-than-adroit hobbyists simply means that the electrolytics have been changed out. Incidentally, the parts costs for paper capacitor replacement is under $10. The time investment is usually four to six hours.

Your new Trans-Oceanic should now have a full complement of good-to-excellent tubes, new electrolytic capacitors, new paper capacitors and such resistors and other parts as were necessary to get the set in full operation. You now have a wonderful radio which should serve you well for many years to come. Congratulations and welcome to the club.

Alignment

There is only one more step necessary to insure that your Trans-Oceanic is operating at full factory specifications: alignment. Any radio receiver is a series of tuned circuits. In general, each circuit passes only one small range of frequencies on to the next stage down the line. You can visualize each circuit as a blank wall with one window in it. When a receiver is being aligned, essentially each blank wall is moved back and forth until its window is aligned with all the other windows before and after it. When everything is aligned, only a signal on one particular frequency can successfully travel from the antenna through all of the windows to reach the loudspeaker and your ears.

Many hobbyists are reluctant to learn how to align radios because it requires special equipment and is thought to be very esoteric. The truth is alignment is as easy as tuning any radio. Today, the "special equipment" required consists of a modest signal generator that can be purchased for $10 or $20 at most ham fests and electronic flea markets. Both the HeathKit and Eico models are particularly reliable. The signal generator is a small tuneable transmitter which sends out a very weak but stable radio signal. Even though signal generators have large dials, they are not to be trusted for accuracy. For accuracy, you also need a shortwave receiver with a digital dial that reads in increments of 5 kHz or less. Even the cheapest SW portable radio with a digital dial will read to 5 kHz. When the Zenith instructions (on the service sheet) say to align on a certain frequency (say 1400 on the AM dial) you simply tune the digital portable to 1400; then tune the signal generator until you hear the generator on the portable. You now *know* that the signal generator is on frequency.

From that point on, you just follow the instructions on the Zenith or the Sams PhotoFact service sheets. All you have to do is follow the instructions carefully and be willing to do each step several times to get things *exactly* peaked up. There are good general discussions of signal generator use and alignment in most of the electronic texts mentioned earlier in this chapter.

There are actually two kinds of alignment necessary to totally peak up the performance of your Trans-Oceanic: Radio Frequency (RF) and Intermediate Frequency (IF) alignment. The IF alignment is the bit more difficult of the two, though still easy. It aligns most of the receiver to the IF of 455 kHz (getting many of the windows aligned *exactly*). It is somewhat unusual for the Trans-Oceanic's IFs to need much realignment. RF alignment results in the dial reading accurately and the rest of the circuitry being "peaked up." After 40 years or so, most of the dial readings on your Trans-Oceanic will be inaccurate. RF alignment is easy and fun, and having the dial read accurately is very important in getting good service, especially on the shortwave bands!

ONE HINT: Quite often, it is impossible to align the 16 meter band on 17.8 MH$_3$ as called for in the Zenith alignment instructions. There just seems to be not enough range in the 17 MHz coil adjustments to get the band aligned correctly. Even repairmen with sophisticated equipment might give up, assuming that various components have aged badly and that alignment of the 16-meter band is just not possible. This is usually not true. If you cannot properly align the dial of the 16 meter band (17.8 MHz), trace the lead wire from the 16-meter band oscillator coil (usually called L20) to the grid of the oscillator tube (usually the 1L6). Move this wire away from all metal chassis parts as far as possible. You should now be able to align the 16-meter band dial perfectly!

If you have performed all of the electronic repair and restoration steps outlined so far and if your alignment is well done, your Trans-Oceanic now operates as well as the day it left the factory. If you are operating one of the 600 Series radios, you now can enjoy what is surely the finest mass-marketed consumer shortwave tube-type receiver made in North America in the post-war era. In fact, because of its bandspread dials which focus on the International Shortwave Broadcast Bands, the later tube-type Trans-Oceanics are better radios for shortwave broadcast listening than all but a handful of large consoles from the early 1940s and the most expensive communications receivers in the late 1950s and 1960s. Your Trans-Oceanic also has better audio quality on shortwave than all but one or two of the most modern and expensive digital dial solid-state radios. Again, congratulations!

Even though you now own a wonderful radio which is performing as well as it did when it came from the factory 40 to 50 years ago, there are very good reasons why you should consider making one modification and fabricating one of several possible accessories for your "new" Trans-Oceanic.

SOME SUGGESTIONS

A Fused and Dampened AC Source

Authorities agree that all tube-type electronic devices should be protected by either slow-blow fuses or very low amperage circuit breakers. They also agree that much of the wear and tear on tube electronic devices is caused by the violent inrush of current which occurs in the first second or so after the set is turned on.

In most radios, the electromagnetic actions of the power transformer and the slow-to-warm-up 6-volt tube filaments at least partially dampen this harsh inrush of current. Since three-way radios like the Trans-Oceanic lack a power transformer, and since 1.5-volt tubes react to current almost instantaneously, the Trans-Oceanic and similar radios are particularly vulnerable to damage by current inrush.

Both the fusing problem and the current inrush problem may be addressed by placing a few new components under the chassis. They may also be addressed by building a simple but effective "extension cord" for tube-type Trans-Oceanics. We have built extension cords using a normal receptacle mounted in a heavy plastic wall outlet box common to home construction. After the 117-volt power comes into the box, one leg goes through a 1-amp circuit breaker. After the breaker, both legs of the power line run through "varistors" before reaching the receptacle and the Trans-Oceanic.

As discussed previously, varistors are small devices which were specifically designed to protect delicate devices from current inrush. The varistors that we use (Allied Electronics Part Number 837-0090) have 120 ohms of resistance at turn-on and cost less than $4 each. As the current flows, their resistance drops. It reaches around 2 ohms after a minute or so. Using these devices will slow down the onset of sound from a

Trans-Oceanic by only five seconds or so, but will give the radio the "soft start" it needs to extend longevity.

Improving Selectivity

When tube-type Trans-Oceanics were being designed and manufactured, the International Shortwave Bands were relatively uncrowded and most unwanted signals were quite weak. During the last three decades, these bands have become extremely crowded with strong signals. This situation often demands a level of electronic selectivity which was not even possible during the heyday of the Trans-Oceanics. Now, the desired signal is often easily tuneable, but it still has an audio whine or howl (a "hetrodyne") in the background. Sometimes this nuisance may be removed by careful tuning, but often this is impossible. Please note that this very same problem exists today even when using the most expensive communications receivers built in the 1950s and 1960s.

KIWA Electronics of Yakima, WA (Appendix II) has developed a series of very high quality IF filters which may be added to any radio with an IF frequency of 455 kHz. Although these filters were originally designed to be installed on solid-state communications receivers, advanced hobbyists have recently begun installing them on tube-type communications receivers and achieving significant improvements. A typical installation in a tube-type receiver was covered in the 1992-1993 edition of the *Proceedings of Fine Tuning*. The filters cost under $50 and may be added under the chassis of the Trans-Oceanic in a completely nonintrusive way. A 4 kHz-wide KIWA filter should remove almost all hetrodyne problems on crowded bands and will allow the inherently excellent audio of the Trans-Oceanic to be enjoyed even under difficult modern conditions.

CONCLUSIONS

It is very unlikely that anyone except a true Trans-Oceanic aficionado will find every suggestion in this chapter useful. Your approach to restoring your Trans-Oceanic to a useful life should be based on your own technical skill level. However, each operation that you undertake should not do anything to permanently degrade the circuitry or appearance of the radio. We are each only custodians of these wonderful radios; preserving them for the future should be our primary goal at all times.

BIBLIOGRAPHY

Bryant, John, *et al. Fine Tunings Proceedings, 1992-1993*. Available for $20.50 plus $4 S&H from: Fine Tuning Special Publications, c/o John H. Bryant, Rt. 5, Box 14, Stillwater, OK 74074.

Carr, Joseph J. *Old Time Radios, Restoration and Repair*. Blue Ridge Summit, PA: Tab Books, 1991.

Marcus, William and Alex Levy. *Practical Radio Servicing*. New York: McGraw-Hill, 1955. [Was also published by at least one other publisher.]

Marcus, William and Alex Levy. *Elements of Radio Servicing*. New York: McGraw-Hill, 1955.

Romney, Ed. *Fixing Up Nice Old Radios*. Self-published, 1990. [Widely available at hobby outlets.]

Chapter 11
PHYSICAL AND ELECTRONIC RESTORATION OF SOLID-STATE MODELS

General Overview

In contrast to the earlier tube models, there is less restoration that can be performed by the owner of a transistorized Trans-Oceanic. This situation is partly due to the different nature of the electronics involved and partly due to the more modern materials used in constructing the cabinetry of these radios. There are some things that a collector *can* do, however, to bring the members of this family of Trans-Oceanics back to useful life and beautiful appearance.

As with the tube models, there are some ethical and philosophical decisions to be made concerning just how much and what type of restoration to perform on the solid-state models. If you have not read the "Zen of Trans-Oceanic Restoration" in Chapter 10, we strongly recommend that you do so before proceeding with any Trans-Oceanic restoration.

A list of tools and materials helpful to the Trans-Oceanic restorer is found in Appendix I; suppliers for the specific products mentioned in the following discussion are presented in Appendix II.

PHYSICAL RESTORATION

The cleaning, repair and restoration of solid-state Trans-Oceanics can most easily be grouped by similarity of models. The Royal 1000 Series and the Royal 3000 Series Trans-Oceanics are very similar radios; in fact, the Royal 3000 can be considered a Royal 1000 with an FM section added to the chassis. As you will note in the section on Parts Commonality (Chapter 12), these two series share about 80% of their cabinet parts and hardware. There is a similar parts commonality pattern between the various versions of the Royal 7000 Series radios and those of the R-7000 Series, although they are electronically totally different radios.

ROYAL 1000 and ROYAL 3000 SERIES TRANS-OCEANICS

If anything other than a cursory cleaning is planned, the first task is to remove the chassis from the cabinet. This is quite easily accomplished (refer to Figure 11-1):

1) Remove the two small and one large knobs from the front panel. They are all friction-fit and may be removed by pulling straight out. Remove the band change knob from the end of the cabinet. *CAREFUL!* Most of these knobs have a set screw which is accessible only from inside the cabinet.
2) Remove the green whip lead-in from the upper antenna screw and unplug the Wavemagnet lead-ins from the upper center of the chassis (black rubber plug).
3) Remove the five hex-head chassis mounting screws. Two are on the right edge of the chassis, two are at the bottom edge and one is *cleverly* hidden at the upper left corner of the chassis, buried far behind the space occupied by the shaft of the tuning knob!
4) Gently shift the chassis in the cabinet to assure yourself that it is loose and then ease it out. As with the tube models, be very careful of the fragile RF coils on the left-hand side of the chassis. Be gentle and have patience: it *will* come out.

Once the chassis is removed, you can make a careful inspection of the entire cabinet and develop a restoration plan.

POLISHED AND BRUSHED CHROME FINISHED METAL PARTS

The metal finishes on the Royal 1000 and Royal 3000 Series radios were a major disaster. Today, it is not readily apparent just what the problem was, but none of these finishes met the test of time well. The least serious problem is the occurrence of almost microscopic-size pitting in both the bright and brushed chrome finish. Although some radios have survived unscathed, we have even found some pitting on brushed and polished chrome on new Zenith spare parts which have not been removed from their boxes since their manufacture in the late 1950s! A far more serious flaw is the tendency of the brushed-chrome plating to bubble and blister. This is a very common occurrence on cabinet tops, upper side panels and the upper portion of the front door. This blistering looks exactly like badly sunburned skin, with individual blisters varying from 1/8-to-1/2-inch diameter. Finally, the gold lacquer on a portion of the brushed chrome front panel of the Royal 1000 Series often seems to have chemically decomposed and may be brushed off like so much dust.

These problems were probably neither foreseeable nor preventable by Zenith. The late 1950s to mid-1960s was an era of rapid development of new metal and plastic technologies and methods of production. These pitting and blistering problems were long-term, slowly developing problems, which would not have shown up in the short cycles of prototype testing and development. Further, Commander McDonald was still alive and very much in personal control of the development of the Royal 1000. Had any of these problems been predictable, we can be sure that things would have been done very differently.

There appear to be no reasonable solutions for most of these metal-related problems. The only possible response to the blistering of the brushed chrome and the deterioration of the gold lacquer is simply to change the affected parts. Luckily, this is rather easy to do, and there are broad areas of commonality across all five variations in the Royal 1000/3000 Series. Please refer to the section on Parts Commonality (Chapter 12) for further discussion of this strategy.

Pitting and small dings in the chrome are best attacked with a good chrome polish. An excellent final step to preserving these metal surfaces is an application of the Novus #1 plastic polish discussed in Chapter 10.

RESTORATION OF BLACK CABINET CLADDING

It is not well known, but the black cladding material on early Royal 1000s was "Genuine Cowhide." If you are not sure whether a particular Royal 1000 is leather-clad, look on the inside surface of the upper part of the front door. If you are one of the fortunate few to own one of these rare leather-clad Royal 1000s, use restoration techniques similar to those discussed in the section covering leather-clad 600 Series Trans-Oceanics in Chapter 10.

The black plastic material covering the exterior surfaces of most Royal 1000s and all Royal 2000s and 3000s is much harder and less flexible than modern sheet vinyl plastic. It is similar in density and feel to sheet PVC. Surfaces of this material respond very well to the two Novus plastic polishes discussed elsewhere.

On a few Trans-Oceanics, this stiff plastic cladding material has "walked" and become badly out of position over the years. It is sometimes possible to peel all but the very bottom layer of this material off the ends of the radio cabinet and reglue it in proper position with contact cement. This is a very chancy maneuver and should be avoided, if at all possible.

With the material problems (chrome and gold enamel), problems of Battery leakage and the structural concerns with many of the combination Wavemagnet handles, finding a nearly flawless Royal 1000/3000 is relatively rare. Most collectors have more working chassis than good cabinets. If you are ever in that situation, you might consider fabricating a replica cabinet from clear plastic. It would be a striking display.

HANDLE AND WAVEROD

In retrospect, the decision to enclose the Waverod in the semi-detachable main cabinet handle was a *very bad* design decision; this design flaw was corrected a decade later when the Royal 7000 Series was designed. Today, it is rare to find a Royal 1000 or Royal 3000 Trans-Oceanic without a cracked carrying handle. The black plastic molding was simply far too thin to withstand the stresses of real world use. Usually the cracks are found in the sides of the handle near the most fragile locking/unlocking end, caused by someone pressing down too vigorously on the handle rather than lifting it.

Whether your handle currently has cracks or not, we recommend the following procedure to reinforce the handle: obtain several ounces of epoxy-based putty at your local hardware store; take the bottom (underneath) portion of the handle off by removing the five small screws; pack the area between the barrel of the Waverod and the upper (main) plastic handle molding more or less solidly with the epoxy-based putty, being careful where it is placed. Allow the epoxy to cure and reassemble the handle. The goal of this project is to reinforce the main plastic molding

by the added thickness of the epoxy putty and to engage the outer barrel of the Waverod as a handle-stiffening device. Take reasonable care in this operation and your Royal 1000/3000 has a much better chance of long-term survival.

RESTORATION OF PLASTIC PARTS

Restoration of the plastic parts of the Royal 1000/3000 Series is best accomplished with Novus #1 and Novus #2 plastic polish (refer to the section "Restoring Plastic Parts" in Chapter 10).

CABINET INTERIOR (including inside of rear door)

Most solid-state Trans-Oceanic owners had at least one "accident" with batteries. This usually resulted in battery acid leaking to the floor and possibly other areas of the interior cabinet surfaces. If the corrosion is limited, restoration may be done without disassembling the cabinet. However, if the splattering is general, you should consider disassembling the cabinet so that a total interior restoration is possible.

In either case, the restoration process is the same. Use 000 and 0000 steel wool to remove *all* of the corrosion down to bare metal. Repaint with silver (aluminum) paint of as near a color match as you can find. When just a few spots are being restored, spray some of the silver (aluminum) paint into a small cup and spot apply with a small artist's brush.

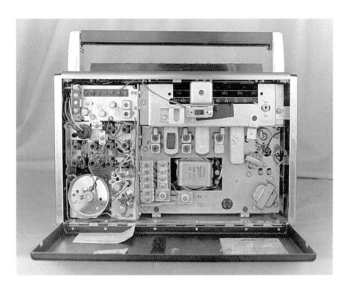

Royal 3000 Series from rear.

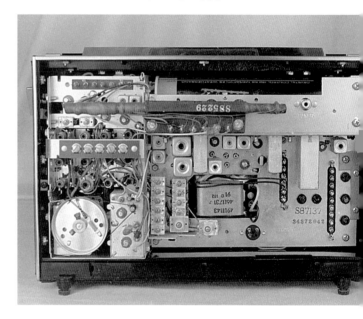

Royal 7000 Series from rear.

Royal 1000 Series from rear.

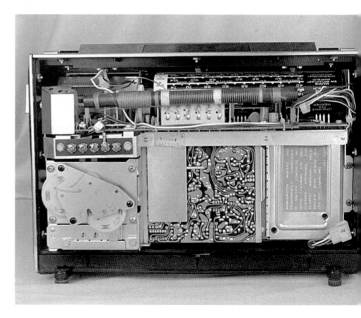

R-7000 Series from rear.

Figure 11-3

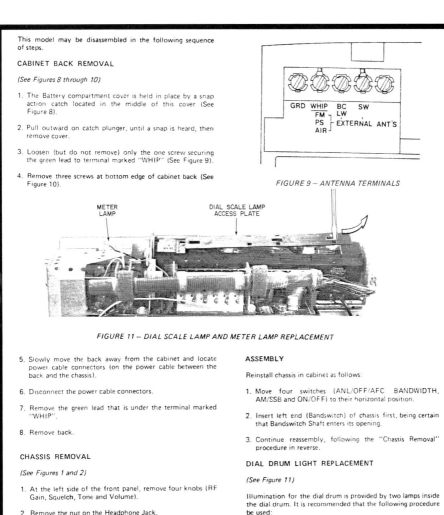

This model may be disassembled in the following sequence of steps.

CABINET BACK REMOVAL

(See Figures 8 through 10)

1. The Battery compartment cover is held in place by a snap action catch located in the middle of this cover (See Figure 8).

2. Pull outward on catch plunger, until a snap is heard, then remove cover.

3. Loosen (but do not remove) only the one screw securing the green lead to terminal marked "WHIP" (See Figure 9).

4. Remove three screws at bottom edge of cabinet back (See Figure 10).

FIGURE 9 — ANTENNA TERMINALS

FIGURE 11 — DIAL SCALE LAMP AND METER LAMP REPLACEMENT

5. Slowly move the back away from the cabinet and locate power cable connectors (on the power cable between the back and the chassis).

6. Disconnect the power cable connectors.

7. Remove the green lead that is under the terminal marked "WHIP".

8. Remove back.

CHASSIS REMOVAL

(See Figures 1 and 2)

1. At the left side of the front panel, remove four knobs (RF Gain, Squelch, Tone and Volume).

2. Remove the nut on the Headphone Jack.

3. Remove two tuning knobs (Coarse and Fine Tuning) at the right side of the front panel.

4. A Bandswitch is located on the right hand end panel. Remove Bandswitch Knob. (It may be desirable to apply pressure to the shank of the knob from just inside the cabinet).

5. Remove four screws mounting chassis to cabinet.

6. To remove chassis, ease the right hand end of the chassis (as viewed from the back) outward first. Then move chassis to the right (so bandswitch will clear its opening) while sliding the chassis out of the cabinet.

7. Disconnect speaker leads.

ASSEMBLY

Reinstall chassis in cabinet as follows:

1. Move four switches (ANL/OFF/AFC BANDWIDTH, AM/SSB and ON/OFF) to their horizontal position.

2. Insert left end (Bandswitch) of chassis first, being certain that Bandswitch Shaft enters its opening.

3. Continue reassembly, following the "Chassis Removal" procedure in reverse.

DIAL DRUM LIGHT REPLACEMENT

(See Figure 11)

Illumination for the dial drum is provided by two lamps inside the dial drum. It is recommended that the following procedure be used:

1. Remove chassis from cabinet.

2. Rotate Bandswitch, and its dial drum, to "AM".

3. Insert screwdriver blade or similar tool, into one of the slots in the plastic dial light bracket and move bracket to the right (facing the rear), disengaging the bracket.

4. Remove dial light bracket approximately half way. CAUTION — Do not break leads.

5. Replace lamps by pulling upwards.

6. Reposition bracket over opening.

7. Move bracket to left to secure.

ROYAL 7000 SERIES
and R-7000 SERIES
TRANS-OCEANICS

Both the design and materials selection of the Royal 7000/R-7000 Series of Trans-Oceanics was much more successful than that of the previous Royal 1000/3000 Series radios. This was probably due both to advances in materials technology and to lessons learned from the Royal 1000/3000 materials failures. Further, receivers from these two series are still really relatively new and are just entering the collecting hobby market. For all of these reasons, a thorough cleaning and minor repair are all that is usually necessary with most Royal 7000/R-7000 Series radios.

These radios respond to the same cleaning techniques discussed for the Royal 1000/3000 Series. However, there is one *caution* worth noting: from time to time, there have been delamination problems with the vinyl cloth on the side panels of the cabinet - the vinyl peels away from the metal side panel. If this happens, it is relatively easy to simply reglue the vinyl. However, *the formulation of the vinyl is such that it reacts violently to normal petroleum-based contact cements!* If normal contact cement is used, the vinyl soon starts to bubble, melt and exude an oily substance. A much better approach would be to use normal white or yellow non-petroleum glues.

Before you begin restoration, you must decide whether to remove the chassis from the cabinet. This is a considerably more complicated operation with the Royal 7000/R-7000 receivers than with the Royal 1000/3000 receivers. Since it is necessary to remove the chassis from the cabinet to do electronic alignment, it is often a good idea to combine cleaning, cosmetic restoration and electronic alignment of these models into one operation. In general, the fewer times the chassis comes out of the cabinet, the better. Few things are so satisfying as bringing a cabinet to its full beauty and then reinstalling a peaked-up chassis whose dial now is exactly accurate on all bands.

CABINET BACK REMOVAL, CHASSIS REMOVAL AND DIAL DRUM LIGHT REMOVAL FOR ROYAL 7000 AND R-7000 SERIES TRANS-OCEANICS

Since the chassis removal and minor repair items for these last two Trans-Oceanics is a bit complex, we have obtained permission from Zenith Corporation to reproduce the instructions from Service Manual RA-95 (Figure 11-3). These descriptions were written for the first version of the R-7000 (Taiwan) Trans-Oceanic; however, all instructions generally apply to all versions of both the Royal 7000 and the R-7000 Series.

ELECTRONIC RESTORATION

The components and the electronic design of solid-state Trans-Oceanics differ a great deal from those of their "hollow-state" ancestors. The electronic restoration of these models naturally differs from that of their tube radio predecessors quite radically as well. One major difference between tube Trans-Oceanics and solid-state models is that professional service should still be available for the newer models at many Zenith dealers or at other electronic repair shops.

Another major difference between the tube and solid-state Trans-Oceanics is the availability of useful service manuals. For tube models, a private company, Sams PhotoFact, as well as Zenith, provided a single schematic circuit diagram per model. Sams continued to provide only one set of schematics for solid-state models; however, Zenith provided a separate Service Manual for each chassis used for each solid-state model. Although the different chassis within a particular model may vary only slightly, we strongly recommend obtaining the correct Zenith Service Manual that matches the chassis in question. Copies of Zenith Service Manuals may still be available at some Zenith dealers and radio clubs. The numerical references for both Sams PhotoFacts and official Zenith Service Manuals are provided as Figure 11-4 and Figure 11-5.

TRANS-OCEANIC MODEL	SAMS PHOTOFACT REFERENCE
Royal 1000	420-14
Royal 1000-D	451-20
Royal 1000-1	none (420-14 similar)
Royal 2000*	560-19
Royal 2000-1*	none (560-19 similar)
Royal 3000	TSM-33
Royal 3000-1	TSM-74
Royal 7000	none (TSM-157 similar ???)
Royal 7000-1	TSM-157
R-7000	none (unique radio)
* Companion radio, not a Trans-Oceanic	

Figure 11-4: SCHEMATIC DIAGRAMS FOR SOLID-STATE TRANS-OCEANICS AVAILABLE FROM SAMS PHOTOFACTS

Repairing Common Problems

It is beyond the scope of this chapter to provide a complete discussion of the servicing and electronic restoration of each of the solid-state Trans-Oceanics. However, there are a number of common problems which occur when restoring any of these models. The following discussion is intended as a general guide to electronic restoration of solid-state models for someone with a moderate amount of electronics experience.

Removing the Chassis

Removing the chassis from any solid-state Trans-Oceanic cabinet is an exercise in patience and manual dexterity. This situation was probably caused by the Zenith's desire to make the Trans-Oceanic as compact as possible. Please refer to the section on Removing the Chassis at the beginning of this chapter. Have patience and be gentle: the chassis really will come out!

First Steps

After removing the set from its cabinet, the first thing that you should do electronically is to carefully clean the chassis and treat all electrical contacts in the bandswitch, slide switches and rotary controls with contact cleaner/enhancer. If you have steady hands, it is also a good idea to remove each transistor from its snap-in mount. [Does not apply to the R-7000 model.] Spray contact/enhancer down into the transistor socket and reseat the transistor gently but firmly in the socket. This cleaning strategy, alone, has returned many Trans-Oceanics to full service. If the set is working relatively well at this point, little else need be done electronically. Since the chassis must be out of the cabinet to perform alignment procedures, you should probably perform an RF alignment before returning the chassis to its cabinet. Most solid-state Trans-Oceanics do not need IF alignment and the RF alignment is usually only necessary to touch up dial accuracy.

Model	Chassis	Printed	Inventory Part No.	Vol/ Pgs	Service Part No.
1000-D	?	8/59	202-1356	7-	
1000-D	?9AT41Z2	8/59	202-1356		
1000-D	?9AT40	5/59	202-1539		
1000-D	9CT41Z2	5/59	202-1539	7-139	
1000-D	9CT41Z2	4/60	202-1539A		
1000-D	9HT41Z2	2/61	202-1790	7-	
1000-D	9HT41Z2	4/61	202-1790A		
1000	?9AT40		202-1301D	7-135	
1000	?9AT41		202-1301D	7-135	
1000	9CT40Z2	8/59	202-1586	7-225	
1000	9CT40Z2	4/60	202-1586A		
1000	9HT40Z2	1/61	202-2137		
1000	9HT40Z2	2/61	202-1789	7-	
1000-1	9HT40Z2	11/63	202-2430		
1000-1	9HT40Z8	6/65	202-2656A		
3000	12KT40Z3	-	202-2112		
3000-1	12KT40Z3	-	202-2422		
3000-1	12KT40Z8	-	202-2645A		
3000-1	12KT40Z8	5/69	202-2645F		
7000	18ZT40Z3	7/69	202-3076A		
7000-1	18ZT40Z	-	202-3371		
7000-1	18ZT40Z	12/71	202-3371A		923-RA-665 19
D7000	500MDR70	12/71	202-3444		
D7000	500MDR70	2/73	202-3444C	-	923-RA-722 39
D7000	500MDR70	2/75	202-3659A	-	923-RA-736 43
D7000	500MDR70	3/76	202-4002	-	923-RA-869 79
R-7000	2WKR70	1/81	-	-	923-RA-957 95
R-7000	2WKR70	9/81	-	-	923-RA-1028 105
R-7000	2WKR70	4/82	-	-	923-RA-1043 111

Figure 11-5: ZENITH SERVICE MANUALS FOR SOLID-STATE TRANS-OCEANICS

Dial Light Replacement

Luckily, dial lights are still available for all models of the Trans-Oceanic, although some are becoming difficult to find.

ROYAL 1000 AND 3000 SERIES:
1) Open the rear door.
2) Remove the removable Wavemagnet (if you are lucky enough to have one).
3) If you can see a good bit of the dial drum, go to Step 4. If you cannot, remove the drum shield, a piece of sheet metal held on with two hex-head screws.
4) Rotate the bandswitch until the brown or tan dial light plate is conveniently exposed.
5) Free the dial light power wire from beneath its metal retaining clip.
6) Remove the two hex-head screws from the dial light plate (one is under the normal position of the dial light power cord).
7) Gently wiggle the dial light plate and lift it.
8) Remove the bulbs, spray contact enhancer on the bulb base and in the socket. Put the bulbs back in and retest them. Often this solves the problem.
9) If the bulbs are burned out, remove, replace and then reverse the process above to reassemble the radio.

ROYAL 7000 SERIES DIAL LIGHT REPLACEMENT:

Please refer to the reproduction of the Zenith Service manual in the section on Physical Restoration. NOTE: Dial lights for all 7000 Series are becoming scarce. There are recent reports that Sylvania's #17 Wedge Light (small version) is a direct replacement.

A "Battery Eliminator" for the Royal 1000-1, Royal 2000-1 and Royal 3000-1

The nomenclature "-1" was added to the Royal 1000, 2000 and 3000 radios when Zenith began producing external "battery eliminator" power cubes in the mid-1960s. These are the now familiar cubes that plug directly into a household wall outlet to power a number of modern electronic devices. They were an innovation in the mid-1960s. When Zenith brought out the new "-1" line, they provided a 3/32" diameter female "input port" jack on the side of the radio and clearly labeled it with "12 Volts DC" stamped on the outside of the cabinet above the input jack. If you are lucky enough to have one of the "-1" models, obtaining a "battery eliminator" to power your Trans-Oceanic is quite easy.

You need a battery eliminator (also known as a "plug-in power supply") which produces 12 volts DC and which will supply at least 200 mA of current. The male jack on the 12v DC end of the eliminator needs to be 3/32" in diameter. This is one of the several standard sizes that are available. Most important of all, *THE TIP OF THE 3/32" MALE PLUG NEEDS TO BE ELECTRICALLY NEGATIVE.*

If your 12v DC, 200 mA battery eliminator has the wrong plug on the end or the wrong polarity, it is relatively easy to install the proper plug, but you must *MAKE SURE THAT THE PLUG TIP IS ELECTRICALLY NEGATIVE.* If you are not electronically adept, a service technician should be able to do this easily for a nominal charge.

Please note that the dial lights will not work from the power cube, but the radio will perform well. This is as the Zenith engineers designed it.

MORE SERIOUS PROBLEMS

If more serious problems exist with your Trans-Oceanic, you should probably take it to an experienced repair technician. The following notes may be useful when dealing with the technician or if you have some experience in electronic repair/restoration yourself.

Capacitors

There are virtually no paper capacitors used in even the earliest transistorized Trans-Oceanics. This change alone made these receivers much more long-lived than their tube ancestors. Very few of the modern capacitors or resistors used in these newer radios fail. If a capacitor does fail, it is usually one of the electrolytics. These electrolytics, however, are the modern highly sealed variety and are themselves very long-lived. Replacing a capacitor in a solid-state Trans-Oceanic is a thankfully rare occurrence.

Rotary Controls

One of the most common failures in radios of this vintage is the on-off switch. This switch is combined with the volume control on all models except the R-7000. Most of these combined volume/on-off controls are located in very densely packed portions of the chassis. It is very difficult to purchase replacements for these controls which are exact matches, both physically and electrically.

There are three other strategies that offer more hope. The first is falling back on a "parts radio" hoping that the switch in that radio was treated less roughly. A second strategy is to wire around the offending switch entirely. The old control should be left in place to continue performing as a volume control. If you wire around the switch, a new small switch may be placed in the circuit for on-off switching. The switch could be mounted on a small aluminum angle clipped under one of the chassis mounting screws on the right rear of the chassis. Of course, the back of the cabinet would have to be opened to switch the radio on or off but the integrity of the cabinet will be preserved. A third, handier approach would be to modify the radio for AC operation via an external "battery eliminator," if it did not come from the factory with that configuration (refer to end of this chapter). The radio would then get its power externally, rather

Zenith	Purpose	SK No.	Note No.
121-44	RF AMP	SK3007A	126
121-46	DRIVER	SK3004	102A
121-47	OUTPUT	SK3004	102A
121-48	OSC	SK3008	160
121-49	MIX	SK3007A	160
121-64	1ST AUD	SK3004	102A
121-73	1ST IF	SK3006	160
121-74	2ND IF	SK3006	160
121-137	1ST IF	SK3006	160
121-138	2ND IF	SK3008	160
121-139	3RD IF	SK3006	160
121-180	1ST IF	SK3006	160
121-181	2ND IF	SK3006	160
121-228	FM RF	SK3006	126
121-230	FM OSMX	SK3006	126
121-294	FM RF	SK3006	126
121-295	FM OSMX	SK3006	126
121-349	RF	SK3006	160
121-350	OSC	SK3246A	160
121-351	MIXER	SK3006	160
121-352	1ST IF	SK3006	160
121-353	1ST IF	SK3006	160
121-373	OUTPUT	SK3004	158
121-374	1ST AUD	SK3004	102A
121-375	DRIVER	SK3004	102A
121-430	1ST AUD	SK3854	123AP
121-430	PRE-DRV	SK3854	123AP
121-433		SK3854	123AP
121-441	DRIVER	SK3466	159
121-613		SK3452	108
121-613	OSC	SK3452	108
121-678	OUTPUT	SK3444	123AP
121-679	OUTPUT	SK3466	159
121-687	IF	SK3039	107
121-687	AM MIX	SK3039	107
121-687	AM OSC	SK3039	107
121-687	AM RF	SK3039	107
121-687	BFO OSC	SK3039	107
121-687	FM OSMX	SK3039	107
121-687	FM RF	SK3039	107
121-687	OSC MIX	SK3039	107
121-692	WB MIX	SK3117	161
121-692	WB RF	SK3117	161
121-701	VOLTREG	SK3444	123AP
121-787		SK3065	222
121-850		SK3854	123AP
121-858		SK3448	132
121-871	AM RF	NONE!	194*
121-872	1ST IF	NONE!	108
121-872	2ND IF	NONE!	108
121-872	3RD IF	NONE!	108
121-872	AM MIX	NONE!	108
121-872	AM OSC	NONE!	108
121-872	BFO OSC	NONE!	108
121-950		SK3246A	229
121-975		SK3854	123AP
121-7141		SK3984?	160?

NOTE: These units are listed as direct replacements for the Zenith parts shown. However, some Trans-Oceanic restorers report a loss of performance when using modern replacements. The modern replacement transistors should only be used when no direct Zenith transistors are available from a "parts radio."

Figure 11-6: TRANSISTOR REPLACEMENT GUIDE SOLID-STATE ZENITH TRANS-OCEANICS

than from the battery box and the on/off switching function could be performed externally. An inexpensive switchable power strip would perform the function nicely. This strategy may seem extreme, but if parts prove unavailable, doing this will provide your Trans-Oceanic with a useful and, we hope, long second life.

Replacement Transistors

Although transistors are more reliable than tubes, they do fail. Unlike tubes, there is no common numbering system for transistors or most integrated circuits; each manufacturer of these parts developed its own numbering system. Zenith, of course, used their own part numbers on the transistors used in the solid-state Trans-Oceanics. Zenith no longer provides these parts to their dealers.

Luckily for us, two companies manufacture large lines of generalized replacement transistors and cross reference those units to all of the major manufacturers, including Zenith. Thomson Consumer Electronics produces a "K" series of replacement transistors and NTE Electronics produces a similar comprehensive "NTE" series of replacements. One, or both, of these series are carried by most major electronic parts houses and catalogs. Each manufacturer lists replacements for almost all Zenith transistors used in the solid-state Trans-Oceanic series. Radio Shack's Archer replacement transistor series only provides replacements for two or three of the most recent Zenith transistors. Figure 11-6 is a comprehensive list of every Zenith transistor type used in any Trans-Oceanic along with the listed "SK" and "NTE" replacement part numbers. It was quite difficult to obtain this information and it is presented here, in full, as an aid to Trans-Oceanic owners and service technicians. We should note, however, that some experienced Trans-Oceanic technicians have not been very satisfied with the performance of these modern replacement transistors. The original Zenith transistors may have been built to tighter or slightly different specifications.

Most of us assume that replacement transistors cost generally what replacement resistors or capacitors cost: pennies per unit. This is not true! In 1994, replacement transistors were running in the $2 to $5 range. This is yet another reason to own a "parts radio" of the exact model and chassis of your "keeper." On the Royal 1000, Royal 3000 and Royal 7000 Series radios, transistors can be swapped just as easily as tubes were in "the old days." Both from a cost and a performance point of view, modern replacements should be used only as a last resort.

Output Transistors

Each Trans-Oceanic except the R-7000 has a "matched pair" of output audio power transistors acting as "Class B" push-pull output amplifier. In early Royal 1000 and 3000 Series radios, these two transistors are the most likely to fail.

As transistors are manufactured, their performance tends to vary within a comparatively broad range. To achieve high quality audio performance, Zenith compared all of their output transistors and coded those of similar performance. Two transistors with quite similar performance (and identical markings) become a "matched pair." Most Royal 1000 and Royal 3000 Series radios used color-coded output transistors. Later models used letters stamped in the top of each of these power transistors. The codes for matched pairs of output transistors are as follows:

Early Royal 1000, Royal 2000, Royal 3000 Series radios output transistors were color-coded either red, white, yellow, green or blue. If one output transistor fails, it should be replaced with one of the same color from a parts radio or both should be replaced with a new matched pair.

The following code was adopted for late model Royal 1000, Royal 2000, Royal 3000 and all Royal 7000 Series radios:

MATCHING GROUP

A	The output transistor matching
B, D	identification is letters stamped in
C, E, G	the top of each transistor. Transis-
F, H, J	tors in any one chassis must be
I, K, M	matched according to the group
L, N, P	chart. Letters can be intermixed
O, Q	but must be from the same group.

It is sometimes possible to use an unmatched pair of output transistors; however, this will usually produce distorted audio which will be most noticeable at low volume levels. This expedient strategy is *not* recommended.

MODIFICATIONS TO SOLID-STATE MODELS

The most important, and easy, modification to early solid-state Trans-Oceanics is to enable the early Royal 1000 and Royal 3000 Trans-Oceanics to be powered by external "battery eliminators" or "power cubes" which plug into normal house current. This is done in a similar way that Zenith engineers made this change to create the Royal 1000-1 and 3000-1 radios. A second modification which is quite easy is one which allows the Royal 1000 and Royal 3000 dial lights to be operated from wall current. The final modification discussed here is to install higher quality IF filters for improved selectivity on the now overcrowded International Shortwave Broadcast Bands.

Installing "Battery Eliminator" Input Ports on Royal 1000 and Royal 3000 Trans-Oceanics

The Royal 1000 and Royal 3000 radios were originally designed to be powered only by their internal battery packs. When the external "battery eliminator" power supplies were developed in the mid-1960s, Zenith engineers designed a battery eliminator for the Trans-Oceanics and mounted a female jack on the right side panel of the cabinet with the words "12 volts DC" stamped in the black plastic near this new jack. This change created the 1000-1 and 3000-1 models.

Even if you do not presently own any of the "-1" models, we suggest setting up the battery eliminator itself with the tip negative and the male plug 3/32" in diameter just as Zenith did originally. You may buy a "-1" model later and will wish to use your battery eliminator with that radio, too. The following description for modifying the radio assumes that you have a battery eliminator with a negative tipped 3/32" male plug which will supply 12v DC. Battery eliminators are available from Radio Shack and most electronic supply houses (refer to battery eliminator comments earlier in this chapter).

To modify a Royal 1000, Royal 2000 or Royal 3000 radio for use with a battery eliminator, you need to obtain the following:

1) A female jack to match the power cube (3/32" diameter).
2) A small section of light aluminum angle (1/2" X 1/2" X 3/4" long works well).
3) A small amount of light gauge, insulated hook-up wire, preferably in two different colors.
4) Two auto electric "butt connectors" for 18-24 gauge wire.

Drill the angle with appropriate-sized holes to accept the female jack and to accept the hex-head chassis mounting screw indicated in Figure 11-7. Solder a 4-inch length of hookup wire to each of the two tabs of the female jack, *MAKING SURE TO NOTE WHICH WIRE WILL BE CARRYING THE NEGATIVE CURRENT FROM THE TIP OF THE BATTERY ELIMINATOR POWER PACK.* Crimp one butt connector to the end of each wire. The other end of the butt connector needs to be distorted slightly so that it will make a positive push fit on the male pins of the chassis mounted battery box plug.

Mount the angle as shown in Figure 11-7 and attach the wires to the studs of the battery jack also as shown. Note that the figure uses yellow wire to carry the negative current. *AGAIN, MAKE SURE THAT THE NEGATIVE CURRENT FROM THE TIP WILL REACH THE PROPER MALE STUD AS SHOWN.*

Plug the battery eliminator into the wall with its jack in the newly installed female jack; turn on the radio and it should work fine.

If you do not feel comfortable doing this, an electronic service technician should be able to perform this hookup for a nominal fee, if you supply the parts. Take this book along.

It is possible to modify the receiver so that it may be powered from either the battery eliminator or the battery pack with automatic switching between the two. This requires a more complex female jack plus some rewiring and is beyond the scope of this book. Please note that this modification, *AS DESCRIBED,* is non-invasive and may be removed quite easily to return the radio to its factory configuration.

Enabling the Dial Lights on Royal 1000 and Royal 3000 Radios

All models of the Royal 1000 and 3000 Series radios, including the "-1" models, only allow the dial lights to function when the battery pack is connected. Leaving the battery pack aboard a vintage Trans-Oceanic is

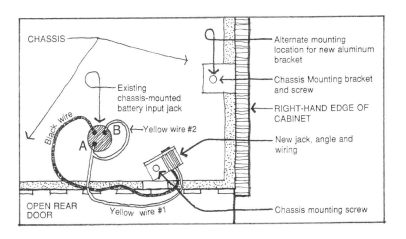

Figure 11-7

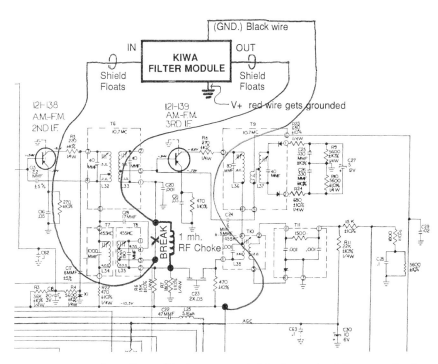

Figure 11-8

an open invitation to battery accidents and serious damage to the radio. *PLEASE DO NOT DO THIS.* Since most modern owners use their Trans-Oceanics only as table models, the following easy modification will enable the dial lights if the radio is a "-1" model or was modified as discussed immediately above.

If the radio is a "-1" model, solder a 4" insulated hookup wire to the (negative) tip terminal of the 12v DC input port jack. Attach a "butt connector" to the other end of the wire and push it over the proper male stud of the chassis-mounted battery box plug. If you have modified your radio to use a battery eliminator as described above, you should solder a jumper wire to the negative power input push connector marked "A" on the Figure 11-7 drawing and use another butt connector on the other end to allow you to plug this new wire into the stud marked B. This will supply 12 volts to the dial light circuit. Change out the existing 1.5v DC dial bulbs for 12v DC bulbs (refer to dial light discussion above). Select 12v DC replacement bulbs which draw the least possible current. Reassemble the radio. The dial lights will now come on when the momentary Dial Light switch is thrown. Do *not* wire around the momentary switch. That action will most likely cause heat to build up inside the plastic dial drum and damage it beyond repair. In extreme cases, a fire may result. If these instructions are followed faithfully, the 12-volt dial lights should provide long years of effective service. This modification may be easily undone to return the receiver to its original configuration.

Adding IF Filters for Improved IF Selectivity

Both the Royal 1000 and Royal 3000 Series radios are good shortwave listening receivers; however, their selectivity is just a bit too broad

for optimum listening on several of the International Shortwave Bands. If there are times when distinct whistles (hetrodynes) bother some of your favorite broadcasts as you listen to your Trans-Oceanic, you should consider installing an accessory IF filter. Currently, there is only one source for these filters: KIWA Electronics. KIWA manufactures a line of filters which may be added to any radio with a 455 kHz IF as all Trans-Oceanics do. The filters are available at various widths from 3.5 kHz to 6.0 kHz. If your primary listening is to International Shortwave Broadcasting, the 4.0 kHz filter is a suitable width for a Trans-Oceanic.

Figure 11-8 is an illustration of the simplicity of adding a KIWA Filter module to, in this case, a Royal 3000. If you are adept at solid state electronics, this modification is quite easy to perform. If you are not, a service technician will be able to perform this task for a modest fee. This modification is also relatively non-invasive and the receiver can be restored to original configuration quite easily.

CONCLUSION

Please note that each operation that you undertake on a Trans-Oceanic, be it restoration or repair, should be undertaken with the long-term preservation of the radio in mind. You should not do anything which will permanently degrade the radio's appearance or circuitry. We are each only custodians of these wonderful receivers. Preserving them for the future should be our primary goal at all times.

Chapter 12

TRANS-OCEANIC
PARTS COMMONALITY

Sources and Strategies

As with most relics of our technological society, no parts specific to any model of Trans-Oceanic are currently being manufactured. Further, none of the specialty mail order houses which serve the vintage radio hobby currently stock any specific Trans-Oceanic parts. This leaves any prospective Trans-Oceanic collector or restorer with only two potential sources of parts: New Old Stock and parts radios.

New Old Stock (N.O.S.): The antique restoration term "New Old Stock" refers to new parts which were manufactured during (or soon after) the manufacture of the radio itself and stockpiled at manufacturers' repair centers and radio service stores. In the case of the Trans-Oceanics, it may still be possible to find longtime Zenith dealers who have some parts in stock. It is currently possible, through *Antique Radio Classified* (see Appendix II) and similar publications, to contact retiring radio and TV repairmen who, on occasion, have a small supply of Trans-Oceanic parts to sell. As the Trans-Oceanic itself fades into history, so too will the sources of N.O.S. parts.

Parts Radios: There will never be any more Zenith Trans-Oceanics in the world than there are right this minute. For that reason alone, the very idea of cannibalizing one radio to bring another back to life falls in an ethical quagmire. Each of us has to deal with this issue individually. In general, though, we should all be *very* reluctant to declare any physically complete Trans-Oceanic as a parts radio. Almost any physically complete Trans-Oceanic can be resurrected with heavy applications of patience and skill. If the restoration of a particular Trans-Oceanic is beyond your current skills, or if it is too time-consuming to contemplate

now, we urge you to carefully store the "basket case" away for a later day. Do not cannibalize *any* set which can be restored at some point in the future.

Even considering the above, the best sources for Trans-Oceanic specific parts are other Trans-Oceanics of the same or similar model. These Trans-Oceanics can be found from the same sources and by the same strategies as previously discussed for finding "collectible" Trans-Oceanics in Chapter 9.

We have found it advisable *never* to discard any part or component of any Trans-Oceanic. What may seem like irreparable trash today may, unpredictably, become very repairable with a new generation of magic "glop." Who could have predicted the resurrection of hardened rubber suction cups by soaking them for a year in Armor-All™? Even parts which are beyond repair can serve very well for trial runs of new restoration techniques. It is often a good idea to practice new techniques before applying them to irreplaceable parts.

Parts Commonality

It is not always possible to obtain a "parts radio" of exactly the same model as the radio being restored. Fortunately, the only Trans-Oceanics which do not share significant parts with other models are the first and last of the breed: the 7G605 Clipper and the R-7000. Each Trans-Oceanic model and its variants share a surprising number of components with other models produced in the same general time frame. Unfortunately, determining parts commonality usually requires having a rather complete collection of Trans-Oceanics at hand. The following studies of Parts Commonality should be of great use.

POST-WAR TUBE MODELS

8G005Y

G500

MODEL 8G005Y AND THE G500

General Comments

The only significant difference between the original post-war Trans-Oceanic, the 8G005Y, and its slightly newer siblings, the 8G005YTZ1 and YTZ2, are modernized power supplies. The newer "YTZ1 or YTZ2" introduced miniature, more modern rectifier tubes in 1948 and 1949. In all other respects, the 8G005Y and the 8G005YTZ1 and YTZ2 are identical radios. They will be referred to, collectively, in the following discussion by the earlier, shorter model number.

The second Trans-Oceanic in this family, the Model G500, was a "transitional" Trans-Oceanic which was produced for only 18 months. It introduced the new thoroughly modern chassis using miniature tubes while maintaining an outward appearance which was almost identical to its predecessor, the 8G005Y. The major visual difference between the two models is that the G500 Wavemagnet carried a bright brass version of the Zenith coat of arms while the earlier 8G005Y offered a much more modest black-on-black "Lighting Bolt Z" in the center of its Wavemagnet.

The chassis of the two models, however, are *radically* different. The chassis 8G005Y is really a significantly upgraded version of the pre-war Trans-Oceanic Clipper. Like its ancestor, the 8G005Y utilized loktal type tubes which were developed before the war for military applications. The G500 introduced the miniature tube-based modern chassis which would, by and large, carry the Trans-Oceanic receiver models to the end of the tube era a decade later.

Cabinets

8G005Y and G500: These cabinets are very similar, however, the 8G005Y provides a horizontal "chin" flip-down panel which conceals a reference and log book. The "chin" of the G500 case is fixed. Further, the mounting systems of the MW Wavemagnet are entirely different in the two models. Finally, the case sides of the 8G005Y are made up of four slightly beveled planes while the side panels of G500 and later Trans-Oceanics are simple flat planes.

In general, it is possible to put a G500 chassis in an 8G005Y case and vice versa. It is also possible to "merge" various pieces of a G500 case with parts of a 8G005Y case to make a single good case. However, the detailed differences in these two cabinets would render the resulting case nearly worthless from an historical or collector's point of view.

Case Front Latch

8G005Y and G500: The case latches of these two models are identical and may be interchanged at will.

SPECIAL NOTE: The *overall* front latch design remained identical from the beginning of the 8G005Y in 1946 until the B600 model was closed out in 1962. All are interchangeable. The case latch of the earlier 8G005Y and G500 are identical to those on later models, *except for the incised lettering*. This lettering is so different, however, that substitution between these

two groups of tube model Trans-Oceanics should be done only as a last resort.

Main Carrying Handle

8G005Y: The fixed handle of this model is unique among the Trans-Oceanics. However, this handle was also used on the two companion radios to the 8G005Y: the 6G001Y Universal portable and the 6G004Y Global portable.

G500, H500 and all 600 Series models: Each of these models has a hinged plastic carrying handle with a slightly different profile; however, the critical dimensions of all of these handles are identical.

Waverod Whip Antenna

The 8G005Y Waverod is unique. All other tube models used identical Waverod antennas and may be exchanged at will.

Wavemagnet

Each model utilizes a unique Wavemagnet. In general, these may not be interchanged.

HOWEVER: the 8G005Y Trans-Oceanic shares an all-black Wavemagnet with the companion 6G001Y and 6G004Y portables mentioned above.

Plastic Front Panels

The plastic main front panel of the 8G005Y is identical to that of the G500. They may be interchanged at will.

Plastic Knobs

8G005Y and G500: Identical, may be interchanged.

Bandswitch Push Buttons

8G005Y and G500: Identical, may be interchanged.

Chassis

8G005Y, 8G005YTZ1, 8G005YTZ2: The chassis of these three variants only differ in the type of rectifier tube used and may be interchanged at will.

8G005Y and G500: The chassis of these two models is RADICALLY different. The 8G005Y utilized loktal type tubes, with only the rectifier tube being changed out for a miniature tube in the 8G005YTZ1. The G500 was the rather quiet introduction of the miniature tube-based modern chassis which would, by and large, carry the Trans-Oceanic receiver models to the end of the tube era a decade later.

H500

600 SERIES

MODEL H500 AND THE 600 SERIES

General Comments

The **H500** and the **600 Series** are nearly identical radios. The chassis and circuitry of each are almost identical, both mechanically and electrically. The general configuration of the H500 cabinet is nearly identical to those of the 600 Series. Although it is not readily apparent, the entire front panel of all 600 Series radios are almost identical with that introduced in the H500. Refer to detailed comments below.

Cabinets

H500 and 600 Series: All of the major cabinet dimensions of the H500 and the 600 Series receivers are identical as are the radii of the corners and rounded edges. However, there are major differences between the two Trans-Oceanic families in placement of large holes as well as differences as to the extent (size) of the rear door. In final analysis, these two receiver cabinets share nothing usable except for some hardware parts with each other or any other model. It is possible, though not advisable, to put a 600 Series chassis in an H500 case and vice versa.

600 Series: All 600 Series cabinets are virtually identical and may be interchanged at will (leather "L" models are an exception, of course, but may be interchanged among themselves).

Case Front Latch

The front latch design remained identical from the beginnings of the H500 in 1951 until the B600 model was closed out in 1962; all are interchangeable. The case latch of the earlier 8G005Y and G500 are identical to those on the later models, except for the incised lettering. This lettering is so different, however, that substitution between these two families of tube model Trans-Oceanics should be done only as a last resort.

Main Carrying Handle

G500, H500 and all 600 Series models: Each of these models sports a hinged plastic carrying handle with a slightly different profile. However, the critical dimensions of all of these handles are all identical. All of these may be interchanged at will, excepting the "L" leather-covered models which have brown rather than black plastic parts.

Waverod Whip Antenna

All tube models except the 8G005Y used identical Waverod antennas which may be interchanged at will.

Wavemagnet

Each model Trans-Oceanic utilizes a unique Wavemagnet. In general, these may not be interchanged.

G500 and H500: These are each unique. However, the easily-lost brass mounting nuts may be interchanged between these two models.

Plastic Front Panels

Although it is not readily apparent, the plastic main front panel of the H500 is almost identical to that of each of the 600 Series models. The only difference is the hole and lettering for the dial light switch on the 600 Series front panel. Some production runs of almost all 600 Series models also mounted a 1/4" female phone jack directly to the lower left corner of the front panel. It should also be noted that H500 front panels seem to have been painted with a distinctly darker gold paint than were the 600 Series panels. This same darker color is also apparent in the varnish covering all brass parts of the H500s.

Plastic Knobs

H500 and 600 Series: Identical, may be interchanged. See gold color note above. The conical golden brass knob inserts are also identical and interchangeable across the H500 and all 600 Series models.

Bandswitch Push Buttons

H500 and 600 Series: Identical and may be interchanged.

Chassis

G500, H500 and all 600 Series models: All of these chassis have identical critical dimensions and mounting hardware. The one slight, but solvable, difference is that the G500 uses a screw-mounted wooden block to secure the upper left portion of the chassis to the case. All other models use a metal upper frame tab screwed directly to the case. This small variation does not prevent complete interchange of all G500, H500 and 600 Series chassis; however, the difference in frequency coverage between the G500 and the H500/600 Series receivers would make interchange of the chassis somewhat confusing.

SOLID-STATE MODELS

ROYAL 1000

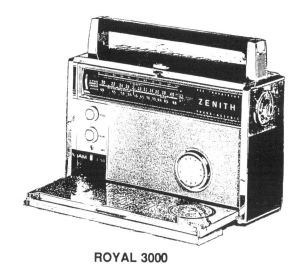

ROYAL 3000

ROYAL 1000, 1000-D, & 1000-1, and ROYAL 3000 & 3000-1

General Weaknesses

The rather lightly constructed black plastic main carrying handle containing the Waverod whip is often found cracked or completely shattered. The chroming of the metal cabinet parts also has tended to blister and pit; a completely unblemished cabinet is rare.

Cabinet

The cabinet used for the Royal 1000 series is largely identical to that used for the later Royal 3000 Series except for a "bulged" rear panel which was added to the Royal 3000 Series cabinet to allow for the thicker chassis of the new AM-FM chassis. A rare exception: the earliest Royal 1000s were covered with black "Genuine Cowhide". Parts thus covered are only interchangeable among themselves.

Handle/Waverod

The somewhat fragile Handle/Waverod is totally interchangeable among all Royal 1000 and Royal 3000 variants.

Cabinet Side Panels

The side panels are totally interchangeable among all Royal 1000 and Royal 3000 variants. However, the left-hand panels of all "-1" models contain a 3/32" phone jack type input port for an outboard 12-volt power supply.

Cabinet Bottom Panels

All cabinet bottom panels are identical in the Royal 1000 and Royal 3000 series.

Cabinet Rear Panels

The model designation of each variant radio in both the Royal 1000 and Royal 3000 series is lettered prominently on the rear panel. Thus, most collectors would hesitate to put a Royal 1000 rear panel on a Royal 1000-D receiver. That aside, the cabinet rear panels of each of the Royal 1000 Series receivers may be interchanged. The rear panels of the Royal 3000 and the Royal 3000-1 are also interchangeable.

Front Panels

Royal 1000 Series: All front panels are identical.
Royal 3000 Series: All front panels use identical.

Dial Escutcheon

One identical rectangular plastic casting is used in all Royal 1000 and Royal 3000 models and may be interchanged at will.

Front Knobs

Royal 1000 Series: The knobs of the three variants are identical, but not interchangeable with any other Trans-Oceanic models.

Royal 3000 Series: The knobs of the two variants are identical, but not interchangeable with any other Trans-Oceanic model.

Bandswitch Knob

Royal 1000 Series: The bandswitch knobs are identical in all three variants EXCEPT that the 1000-D knob has notation for the additional long wave band (see comments for the Royal 3000 Series below).

Royal 3000 Series: The bandswitch knob on both variants is identical. Further, the Royal 3000 Series bandswitch knobs are identical to the Royal 1000 Series knobs, except the additional notation of the FM band.

Front Doors

Royal 1000 Series: The front doors, both top and bottom elements, are identical in all variants.

Royal 3000 Series: The front doors, both top and bottom elements, are identical in all variants.

SPECIAL NOTE: It appears that the upper front door panels of the Royal 1000 and Royal 3000 models are identical. The lower half of the hinged front door of the two series is quite different between the two series.

Chassis

Royal 1000 Series: The three chassis are physically interchangeable. Thought should be given, however, to the reduced resale value of a mismatched case and chassis (i.e., a Royal 1000-D chassis in a Royal 1000 case). See note below for Royal 1000-1.

Royal 3000 Series: The two chassis are interchangeable. The difference between the Royal 3000 and the 3000-1 chassis is the introduction of an external 12v power supply coming through a 3/32" female phone jack in the left side panel of the case. Refer to the modification section of Chapter 11 for instructions to modify a 3000 chassis to emulate this aspect of the Royal 3000-1 (also applies to Royal 1000-1).

Battery Boxes

The earliest Royal 1000s and 1000-Ds were built with a rigid, somewhat brittle battery box permanently attached to the inner surface of the rear case door. This rigid box caused numerous problems over the years. By the early 1960s, Zenith switched to flexible soft plastic boxes on new production runs of Trans-Oceanics. These boxes fit tightly into the cavity between the chassis and the rear cabinet door. All Zenith dealers were also supplied with battery box changeout kits to retrofit the new, nearly unbreakable boxes into the early models. These soft plastic boxes are interchangeable among all Royal 1000 and Royal 3000 variants.

ROYAL 7000

ROYAL 7000 SERIES
ROYAL 7000Y, ROYAL 7000Y-1,
AND ROYAL D7000Y

NOTE: The Trans-Oceanic Model R-7000 (1979-1981) is *not* part of the Royal 7000 Series. See separate section below.

All of the Royal 7000 Series cases and hardware appear to be physically interchangeable. However, with the introduction of the D7000Y, the inner surface of the front cabinet door changed from a brushed chrome finish to a black leatherette finish and the raised logo on the outside of the front door changed completely. The D7000Y also introduced a few subtle color changes in other small detailed areas. However, it appears that all case parts and the case itself are directly interchangeable among the various Royal 7000 Series variants. The Royal 7000 series shares no parts with the previous Royal 1000 and Royal 3000 series.

R-7000

The R-7000 is not part of the Royal 7000 Series. It is a completely different radio and shares few, if any, significant parts with the true Royal 7000 Series.

R-7000

ENDPIECE

As we began research on our next Zenith book we discovered several pieces of important information on the Trans-Oceanic series. Although the printing of this book was imminent, we felt these items were important enough that they must be included in this book. Peter Schiffer of Schiffer Publishing agreed, and with his help, we were able to complete the Trans-Oceanic story with this endpiece.

Trans-Oceanic Development

Part of the additional information, which came primarily from the private files of Zenith founder Eugene F. McDonald, Jr., verified in great detail the story of Trans-Oceanic development as set forth in the material related to endnote 59, Chapter 2. Commander McDonald's radiogram of August 2, 1939, sent from McGreagor Bay, was the single-most important act to initiate research that led to the first Trans-Oceanic. McDonald's files on the original Trans-Oceanic run from 1939 through 1946 and credit engineers G.E. Gustafson and G.O. Striker, along with McDonald, as the originators of the Trans-Oceanic. The files further credit engineers Brown, Emde, Sharp and Thompson as having major roles in refining the design.

The Commander had originally hoped to release the Trans-Oceanic by Christmas 1939, but development of the first shortwave portable radio proved much more difficult than he initially anticipated. Among the problems to be solved were antenna coupling, pickup of external noise, whether or not to use the Radiorgan, poor performance on shortwave, breakup of shortwave with increase of volume, hiss (termed "the waterfall effect" by McDonald) and placement of the Shortwave magnet. In a November 7, 1939, letter to his friend, Jess B. Hawley, McDonald stated, "I am not going to put this on the market until it is a performing fool and there are many 'bugs' to be ironed out."

On October 16, 1941, however, McDonald sent the following memo to Mr. Gustafson: "A year ago last Christmas I looked forward to getting the Deluxe portable [McDonald's term for the 7G605]. I again looked for it last Christmas. I am back for a third time. Am I going to get it for this Christmas? When do your plans call for these to start coming off the line?" Gustafson assured McDonald that the Trans-Oceanic would be in production by December 15, and it was.

A Second Military Model

When we found the production figures above, we discovered there were two military models. Most serious Trans-Oceanic enthusiasts are familiar with the upgraded H500 military Model R-520/URR discussed in this book. The Zenith production list designated this radio "JJ500" and shows the production of 7,218 units in 1952 and 1953. We and our sources within Zenith were very surprised to see the 1956 production figures list the production of 2,973 units noted as "JB600". Thanks to well known Zenith collector Bill Wade, we were able to see photographs of Roger Smith's JB600. It is a 600 series unit covered in green oilcloth bearing an official military tag numbering it as R520A/URR. If you are lucky enough to own this model, you have the rarest of all Trans-Oceanics.

Production Figures

While researching important mileposts in Zenith corporate history, we finally found partial production figures for the Trans-Oceanic. We discovered that Zenith staff had made a numerical analysis of all Zenith products produced prior to July 1965 in an effort to find and celebrate the fifty millionth Zenith Product. The data sheets indicate that 1,029,744 Trans-Oceanics had been produced by May 31, 1965, as follows:

YEAR	TRANS-OCEANIC RADIO MODEL NO.	QUANTITY
1941	7G605	1,257
1942	7G605	33,479
1945	7G605	16
1946	7G605	3
	8G005	28,086
1947	8G005	50,430
1948	8G005	42,034
1949	8G005	10,509
	G500	16,244
1950	8G005	1
	G500	51,634
1951	8G005	16
	G500	21,803
	H500	73,889
1952	JJ500	1,464
	H500	99,256
1953	JJ500	5,754
	H500	71,890
	L600	5,473
1954	H500	509
	L600	41,545
	R600	16,632
1955	R600	12,053
	T&Y600	36,951
1956	JB600	2,973
	600	53,124
1957	600	42,783
	R1000 & R1000D	3,237
1958	600	17,449
	R1000 & R1000D	32,456
1959	600	16,761
	R1000 & R1000D	24,282
1960	600	13,545
	R1000 & R1000D	33,259
1961	600	7,679
	R1000 & R1000D	26,741
1962	600	5,038
	R1000 & R1000D	27,765
	R3000	6,159
1963	600	66
	R1000 & R1000D	8,323
	R3000	31,707
1964	R1000 & R1000D	6,840
	R3000	35,852
1965 (5 mos)	R1000 & R1000D	2,818
	R3000	9,959

TOTALS	
Total Tube Trans-Oceanics	780,346
7G605	34,755
8G005Y	110,567
G500	89,681
JJ500	7,218
H500	245,544
TJB600	2,973
600 Series	269,099
Total R1000/R1000D (to 31 May 1965)	165,721
Total R3000 (to 31 May 1965)	83,677

ENDNOTES

Chapter 1: THE PORTABLE RADIO

1. An excellent discussion of the regenerative circuit is found in "Some Recent Developments of Regenerative Circuits" in *Proceedings of the Institute of Radio Engineers* 10(4):244-260; August 1922. An account of the many contributions of Armstrong to early radio is found in "Creativity in Radio: Contributions of Major Edwin H. Armstrong," in *Journal of Engineering Education* 45(2); October 1954. The famous regenerative circuit patent wars of Armstrong and De Forest are dealt with fully in: Lewis, Tom. *Empire of the Air: The Men Who Made Radio*. New York: HarperCollins Publishers, 1991.
2. Schiffer, Michael. *The Portable Radio in American Life*. Tucson: The University of Arizona Press, 1992, 52.
3. Schiffer, 63-74.
4. Hattwick, Richard. "Eugene F. McDonald, Jr., of Zenith Radio Corporation." Macomb, Illinois: Illinois Business Hall of Fame Studies of Business Leadership. Western Illinois University, Undated, 4.
5. A greatly compressed account of the development of the early Zenith radios is presented in *The Zenith Story*, a 1955 publication of the Zenith Electronics Corporation.
6. Schiffer, 63-86.
7. Schiffer, 67.
8. Schiffer, 63-86.
9. Lorenzen, Howard O. Telephone interview with author, November 24, 1992. Lorenzen worked for several months on the early prototype conversion of the 5G401 chassis.
10. Schiffer, 197.
11. Zenith Archives.

Chapter 2: THE COMMANDER AND THE TRANS-OCEANIC

1. "Commander McDonald of Zenith." *Fortune Magazine* (June 1945): 141-143, 209-210, 212, 214, 216.
2. Calibraro, Jim. "McDonald of Chicago." *Chicago Omnibus* (June 1967): 45-47.
3. *Who's Who in America*, 1959-1960, lists McDonald's birth year as 1890; however, a wide variety of references, including McDonald's obituary notice, place the date at 1886.
4. Hampson, Philip. "The Road to Success: A Sketch of Eugene F. McDonald, Jr., President of Zenith Radio Corp." *Chicago Tribune* (June 6, 1953).
5. *Fortune Magazine* (June 1945).
6. Kinney, Eugene M. Interview with author, Chicago, Illinois, June 1993. Kinney is Commander McDonald's nephew and a former senior vice president and member of the board of Zenith.
7. *Fortune Magazine* (June 1945).
8. Hampson, *Chicago Tribune* (June 6, 1953).
9. *Fortune Magazine* (June 1945).
10. McDonald in later years turned this handicap into profit when the McDonald-inspired Zenith miniature hearing aid, because of its design and low price, destroyed the competition in the market place and established Zenith as the worldwide hearing aid leader.
11. Hampson, *Chicago Tribune* (June 6, 1953).
12. Hampson, *Chicago Tribune* (June 6, 1953).
13. *Fortune Magazine* (June 1945).
14. Pennmann, Jack. "The 'Good-Bad Boy' of Radio." *Future Magazine* (June 1939): 26-32.
15. *Fortune Magazine* (June 1945).
16. Calibraro, *Chicago Omnibus* (June 1967).
17. Hampson, *Chicago Tribune* (June 6, 1953).
18. Brown, Elger. "Chicago Profiles." *Chicago Tribune Pictorial Review* (February 14, 1954): 15.
19. Pennmann, *Future Magazine* (June 1939).
20. Hampson, *Chicago Tribune* (June 6, 1953). The authors verified much of the information dealing with McDonald's early life (as reported in the popular press) through a Christmas letter Commander McDonald wrote to long-time friend Herman Staebler on December 18, 1942. This letter, from the Commander's personal files, detailed much of McDonald's early life as well as the story of the development of the Trans-Oceanic (see note 28 below).
21. *Fortune Magazine* (June 1945).
22. The Zenith Archives contain a letter written on April 4, 1939, by R.H.G. Mathews (Ralph Mathews) to Commander McDonald. McDonald had asked for Mathews' opinion of the actual date of the beginning of Zenith radio manufacturing so that the company's longevity could be used in advertising. In this letter, written on "Ford, Browne & Mathews Advertising" stationery, Mathews recalled the early history of Zenith:

"In 1912, I built my first amateur radio station under the call of 9IK. While experimenting with various forms of rotary spark gaps, I discovered that a sawtooth type of disc had certain technical advantages and while going to Lane Technical High School during the years of 1913 and 1914, I built some aluminum sawtooth spark gaps of this type in the Lane Technical shops, which I found very successful. This gap had a distinctive note and was the subject of considerable comment over the air when I was working with other amateurs. When I graduated from high school in 1914, my amateur radio station had achieved sufficient prominence that various amateurs were asking me to build equipment for them similar to that which I was using in my own station. This seemed to be a pretty good way of making some additional money to help finance my expenses through college. I was working during the summer as a shipboard radio operator to help pay these expenses, but at that time, shipboard operators got paid something like $25 to $30 a month, which was far from sufficient. From 1915 until the war, I made most of my college and personal expenses by building and selling rotary gaps, radio receivers of various kinds and other equipment for amateur purposes. During this time, my station call was changed to 9ZN, and with the increased prominence of the station call, the products became known as 9ZN spark gaps or 9ZN receivers and from this, it was a short step to the 'Z'nith' name.

"During my last year of college, I went into naval service, which of course put a stop to my commercial and amateur radio activities during the period of the war. During the war, of course, I was continually engaged in communications work in the Naval Radio Laboratory in Great Lakes and elsewhere. I did not receive my release from active duty until almost a year after the Armistice, but during this period I resumed the building of equipment for the amateur use on a much larger scale with Karl Hassel, M.B. Lowe and Larry Dutton helping me out on it. The rest of the story you already know.

"To the best of my knowledge, there is no manufacturer now in business who dates earlier than this.

"Additional information which may be of interest is the fact that the Chicago Radio Laboratory, which was a name under which I operated, held the No. 2 Armstrong regenerative circuit license and also in 1919 held a license from Dr. Lee DeForest under various patents of his relating to the use of honeycomb coils, Ultraudion Oscillators and so forth."

R.H.G. Mathews stayed only a few years with Zenith after the incorporation.

Karl Hassel, the other partner, became an amateur radio operator in 1912. According to a Zenith Radio Corporation Press Release dated July 7, 1975, and titled "Radio Pioneer Karl Hassel Dies," Hassel attended Westminster College and matriculated from the University of Pittsburgh in 1915. He entered the Navy in 1915 and met Mathews at Great Lakes Naval Training Station. In 1918 they formed the Chicago Radio Laboratory at the site of the old Edgewater Beach Hotel. Their factory, as the story has passed down, was a kitchen table and consisted of a soldering iron that had to be heated over a gas stove. Initially they produced only one ham radio station a week.

Karl Hassel stayed with Zenith Radio Corporation in various capacities for 55 years.

23. Hampson, *Chicago Tribune* (June 6, 1953).

24. Hampson, *Chicago Tribune* (June 6, 1953).

25. *Fortune Magazine* (June 1945).

26. Archer, Gleason. *Big Business and Radio*. New York: The American Historical Company, Inc., 1939, 17.

27. Zenith Radio Corporation. "Commander Eugene F. McDonald, Jr." *Electronics Digest* (May/June 1971): 16-20.

28. Of particular interest is the Staebler letter dated December 18, 1942, in which Commander McDonald recounted in some detail his own involvement in the development of Zenith Radio. Staebler had introduced McDonald to Mathews and Hassel in 1920 and in this letter the Commander wanted to tell Staebler the result of the introduction. After a brief introduction and the detailing of the development of the Trans-Oceanic (which McDonald was sending to Staebler as a belated "thank you" gift), the Commander wrote of the early history of Zenith after stating, "I'm wondering if you have clear in your mind what really happened back in 1920 when you introduced me to Mathews and Hassel. It's clear in my mind, hence I'm going to refresh your memory." After recounting the Layton garage story, the letter continued:

"I then started an investigation with my patent attorney. He advised me not to go into radio unless I could get an Armstrong license. I investigated this and found that Armstrong would issue no more licenses; therefore I had somewhat given up the idea when one day you told me at the club that you knew a couple of youngsters out on the North Side that were building radios. I went out with you and you introduced me to Mathews and Hassel--the two owners of the company. I asked them if I could buy a set, and they asked me whether or not I had the money with me. I told them that I could give them a check immediately, and they decided they could deliver a radio to me. I asked them to install it in the club so that I could get familiar with the operation of it and later on I'd move it to my yacht.

"I then asked them if I might see their factory. As we walked through it I noticed that the only piece of machinery in there was one electric drill press--the rest of it was all hand work. They were having their panels engraved outside and in reality were just doing an assembly job and turning out about one or two sets a day. I asked them how business was. Mathews showed me a stack of letters containing orders over a foot high. I asked them why they didn't put machinery in and build in quantities. He said they didn't have the money. It was then I noticed engraved on the front of their panels 'Licensed Under Armstrong Patents' and they assured me they had a license, but I found out unfortunately that the license ran to a co-partnership and Armstrong not issuing any more, would not re-issue and make it to a corporation. In spite of all three handicaps, however, I went into a conference with these two boys and their attorney--Mr. Irving Harriott. I told the attorney representing them that he could represent me, too, and that man has become one of my closest friends and is General Counsel for the Zenith Radio Corporation. We are today by far his biggest account.

"I financed these boys only as General Manager and had to depend upon them as a Co-partnership and deal honestly with me. I made a contract with them that they were to get $100 a week apiece for ten years.......The little company that used to turn out one or two radios a day, when you introduced me to them, last week turned out and delivered to the Government over $1,100,000 worth of radios in two days.

"That's the romance of Zenith and the consequences of your introducing me to the two boys! I though you'd just like to have a record of this for your files...Thanks for the favor you did in introducing me."

On December 26, 1942, Staebler wrote a thank you letter to the Commander which included the following:

"I'm going to give you the rest of the story. How I happened to stop at Z-9 station. One of our mechanics at the garage at Ann Arbor in 1918 and 1919 was a radio bug, he had a transmitter and receiving set, one of two in Ann Arbor. He made me one with DeForest Honey Comb coils, verniers, head phones, galena crystal, etc. It was the last word, a beauty, still is.

"We would go to the lake to my cottage where we had a good aerial and little disturbance or interference. The station we talked to, received from us by key and voice was Z-9. They even phoned messages to my friends that I gave them over the air.

"Of course when Bahr and I went by their station, which then

was just north of Edgewater Beach Hotel, I wanted to stop and meet the boys which we did..."

29. Original Patent for Trademark 164,341. Application filed April 24, 1922, as Serial No. 162,805, registered February 20, 1923.

30. Boucheron, Pierre. "Seagoing Industrialist." *The Rudder* (March 1936): 24-27.

31. *Electronics Digest* (May/June 1971): 16-19.

32. McDonald, Eugene. "The Arctic Inspires a New Market: A Romantic Tale of How the Far North Formed the Experimental Laboratory for the Development of Zenith Farm Radio." Unpublished and undated eight page manuscript in the Zenith Archives. An edited version of this manuscript was published in *Executive Service Bulletin*, Volume 14, No. 11, November 1936. This publication was produced by the Policyholder Services Group Insurance Division of the Metropolitan Life Insurance Company.

33. Mackey, David. "The National Association of Broadcasters--Its First Twenty Years." Ann Arbor, Michigan: University Microfilms, 1959, 17-26.

34. Grace, Al. "Man with a Golden Touch." *Everyday Magazine*. St. Louis Post Dispatch (July 8, 1944).

35. Archer, 1939, 271.

36. An excellent appraisal of the radio conferences and political events that led up to this case may be found in: Archer, Gleason. *History of Radio to 1926*. New York: The American Historical Society, Inc., 1938, 318-359.

37. Archer, 1939, 271.

38. *Fortune Magazine* (June 1945).

39. Mackey, 17-26.

40. Mackey, 25.

41. McDermott, William. "Gene McDonald Pioneers Again." *Forbes* (June 15, 1944): 12-14.

42. This was the first time shortwave equipment had been used in airplanes "in mission" and was McDonald's idea. As the aircraft surveyed the more than 300,000 square miles of the Arctic study area, they stayed in constant contact with the *Bowdoin*, further demonstrating the usefulness of shortwave radio to the military. A brief account of this aspect of the 1925 National Geographic Arctic Research Expedition is found in the Zenith Archives.

43. McDermott, *Forbes* (June 15, 1944): 12-14.

44. McDermott, *Forbes* (June 14, 1944): 12-14.

45. Leitzell, Ted. "Two-Fisted Dreamer: The Story of Fighting Gene McDonald." *The American Weekly* (July 13, 1947): 16-17 (see also note 32). Leitzell was the first director of public relations at Zenith and was hired directly by Commander McDonald.

46. Refer to endnote 32.

47. The story, "Riches from the Wind," by Arthur Van Vlissingen, is found in *Popular Science Monthly*, May 1938, and details the development of the wind generator/battery charger by the Albers brothers.

48. McDonald, 7.

49. *Fortune Magazine* (June 1945).

50. According to written information from Robert Adler of Zenith, "WEFM (for 'E.F. McDonald') was built by a group of Zenith engineers led by Bill Phillips and Stan Jones. It broadcast classical music throughout World War II and for several years afterwards. It had a fine record library and enough power to be heard in southern Wisconsin and northern Indiana."

51. *Fortune Magazine* (June 1945). McDonald is credited with twenty-five patents and three design patents between February 27, 1917, and July 15, 1958. Copies of the patent documents are found in the Zenith Archives. Among McDonald's patents are many of the major components of what was to become the Trans-Oceanic receiver.

52. McDermott, William. "Irrepressible Gene McDonald." *The Reader's Digest* (reprint, July 1944): 1-4.

53. Commander McDonald was a Fellow of the Royal Geographic Society of London and a member of the Explorers Club of New York. From an obituary notice in the Zenith Archives.

54. Boucheron, 25-26. After Dr. Ritter's death, Dore Strauch published the account of their search for Utopia in *Satan Came to Eden*; Commander McDonald wrote the introduction for the book.

55. Gustafson, G.E. "The Mizpah-Amateur Transmitter." *Radio News* (April 1938): 44-45, 58-59.

56. Kinney, Eugene. Interview with the authors, Chicago, Illinois, June 1993 (Kinney is McDonald's nephew). McDonald's interest in the Radio Nurse was prompted by the 1932 Lindbergh kidnapping.

57. U.S. Patent No. 2,164,251, filed May 4, 1939, and granted June 27, 1939. Text and drawings. McDonald personally patented a series of similar antennas between 1939 and 1941, including other forms of the Wavemagnet, Patents 2,237,260 and 2,329,634.

58. Memo from Gene McDonald to Steve Healey dated November 9, 1951. Zenith Archives.
59. Commander McDonald to Herman Staebler, December 18, 1942. McDonald Private Files.
60. Memo of thanks from Commander McDonald to "Mr. Gustafson, Mr. Passow, Mr. Striker, Mr. Emde, and all the other men who worked on the short-wave Deluxe portable." January 16, 1942. McDonald Personal Files. According to written information from Robert Adler (see note on Adler below) of Zenith, "The engineering of the Trans-Oceanic was finished by George O. Striker, a young European engineer who joined Zenith in the summer of 1940." Striker died in the summer of 1992 in his native Hungary.

   Adler, a retired still-working vice president of research at Zenith, is the "father of remote control," having patented the first system for Zenith. He holds over 180 patents.
61. Lorenzen, Howard O. Telephone interview with author, November 24, 1992. Lorenzen worked at Zenith in 1939 and 1940. He started his electronics career as a design engineer at Colonial Radio Corporation in Buffalo. After leaving Zenith, Lorenzen went to the Naval Research Laboratory and in 1971 was named head of the Space Systems Division where he was responsible for the Navy's entire satellite program. He retired from NRL in 1973.
62. "Test New Short Wave Radios on Planes, Trains." *Chicago Daily Tribune* (October 4, 1941): 14.
63. This October date was inferred from a variety of documents in the Zenith Archives, including internal memos from sales and production. The first run of chassis was 10,000 units. It is obvious that no real quantities of finished radios were produced until December 1941. Commander Mcdonald had enough on hand by the week before Christmas to send complimentary Clippers to over fifty of his friends and business associates.
64. The name "Trans-Ocean Clipper" appeared only in very early print advertising. Internal memos of the period simply refer to the radio as the "deluxe shortwave portable." Sales bulletins from the Sales Promotion Department of Zenith, however, use "Trans-Ocean" throughout the three-and-one-half-months of Clipper production. A Davega City Radio ad in the January 23, 1942, New York Times uses "Trans-Oceanic" and in a January 24 letter to E.G. Herrmann, Commander McDonald states, "...we have now released 30,000 of the Trans-Oceanic Portables."
65. Zenith Radio "Unusually Important Message" to distributors, December 17, 1941. Zenith Archives. This memo also detailed the many features of the new portable and was attached to a 26-page packet of order forms and promotional material. The order forms, with prices effective December 17, 1941, list the Model 7G605, with battery, at $75. The cost to the dealer was $36.80. These prices were F.O.B. factory, Chicago.

   Commander McDonald was not pleased when in January 1942 he discovered that the West Cost "fellows" (distributors) had raised the price of the Clipper by $15, ostensibly to cover shipping but actually to make extra profit. In a January 26, 1942, memo to Mr. Nance (Sales), he stated, "The raising of the price $15 just burns me up and it particulary burns me up because that profit does not go to the dealer but into the pocket of the distributor." The Commander responded by changing the advertising copy to include the price of $75 and in small lettering "slightly higher West of the Rockies to cover transportation."

   The promotional photographs of the Clipper included in the December 17 packet pictured the receiver with the sailboat grill. A December 17, 1941, memo from Commander McDonald to a Mr. Terwilliger (Sales Promotion Manager) stated, "Get busy on your literature to follow and make new pictures of the Deluxe with the four motor bomber on it. Don't hold anything up though in order to make this change. Make it on that which is not already printed." The bomber grill, therefore, must have appeared on all but the very first Trans-Oceanics.
66. Copies of all the letters are in McDonald's Personal Files. Among the recipients were the Presidents of Ford, Chrysler, Nash-Kelvinator and Buick Motors and an assortment of engineers and technical people. The Commander sent "copies of my latest 'baby'" to many personal friends and the Chairman of the FCC.
67. This information was compiled from many early January 1942 memos in Commander McDonald's Personal Files.
68. Memo to all Zenith Distributors from the Sales Promotion Department dated February 2, 1942: "Three Exclusive Windows on the Trans-Ocean Clipper Take Chicago by Storm." An internal memo from Mr. Terwilliger, Sales Promotion Manager, to Commander McDonald dated January 29, 1942, details the pricing of the display pieces. Dealers were urged to display so that the Wavemagnets were on the window and the "fishpole" antenna (the Commander's term for the waverod or whip antenna) up. McDonald Personal Files.
69. Letter from Commander McDonald to E.G. Herrmann, January 24, 1942. McDonald Personal Files.
70. Letter from McDonald to Fred Zeder, January 12, 1942. McDonald Personal Files.
71. This test was devised and managed by Commander McDonald. McDonald's Personal Files contain many memos referring to this operation, as well as the airline schedules between Miami and Chicago and the arrangements for Zenith employees to meet the portable at the Chicago airport. He requested that the radio be returned to him after testing since he wanted to keep it after all it had been through.
72. Memo "To All Zenith Distributors" dated April 22, 1942. Subject: Last Radio Produced. McDonald Personal Files. This memo contains information dealing with the last Trans-Oceanic produced and is attached to four publicity photographs: "Zenith Sweeps Clean to Lick the Axis," "Zenith's President Completes Last of Famous Portables," "Zenith President Puts 'Last Portable Radio' Through Its Paces," and "Last Zenith Portable Sees Radio Industry 100% for War." An article dealing with the last portable produced, "Zenith President Completes Last of Famous Clipper Portables," is found in the *Zenith Radiorgan* Vol. 1, No. 1 (September 1942).
73. A variety of documents exist in the Zenith Archives that validate the 100,000 unfilled order figure, including letters from McDonald to many of his friends and business associates. Factory production figures in the Zenith Archives indicate that there were three runs of the Clipper: 10,000, 20,000 and 5,000, in that order.
74. Early survey figures are in a memo to Mr. Allen from "PJJ," January 27, 1942. McDonald Personal Files. Other memos scattered through McDonald's 1942 Private Files attest to the decidedly upper class clientele who purchased the Clipper in 1942.

Chapter 3: THE TRANS-OCEANIC IN WAR AND PEACE

1. Virtually from the first day of production, McDonald realized the potential of the Trans-Oceanic as a valuable radio for the inevitable blackouts and power shortages which would occur at home during wartime. On January 16, 1942, McDonald asked Terwilliger of the Sales Promotion Department to prepare a news release detailing the use of the Trans-Oceanic "for civilian defense workers' special use" and mail it to all Zenith distributors. At the same time the release was forwarded to 2,000 daily newspapers. The release prompted Peter Williamson to write "Radio for Blackouts and Power Line Failures" in the May 1942 issue of *Radio News*.
2. Letter from Allen Agers to Zenith Radio Corporation. December 25, 1953. Zenith Archives.
3. Kinney, Eugene M. Interview with authors, Chicago, Illinois, June 1993. Several memos in McDonald's personal files allude to remaining portables saved for special presentations, including a memo from Commander McDonald dated "12-4-42" (eight months after shutdown of the Trans-Oceanic assembly line): "Let me see what our inventory is on short-wave portables now. I'm speaking of my stock and Mr. Nance's. McDonald." The response was that there were 540 [sic] remaining, 298 in Commander McDonald's stockpile and 248 in Mr. Nance's.
4. Letter from B.A. Jacobson to Zenith Radio Corporation. Zenith Archives. This endorsement appeared in a major undated endorsement booklet that was made to emphasize the versatility of the Trans-Oceanic for yachting purposes.
5. Letter from Daniel MacLea, Jr., to Zenith Radio Corporation. December 17, 1954. Zenith Archives.
6. Letter from Lt. Peter Betts, Lt. William Edwards and Lt. Joseph Roy to Zenith Radio Corporation. July 18, 1951. Zenith Archives.
7. Letter from Konstantin Kalser, President and Executive Producer, Marathon TV Newsreel, New York, to Ted Leitzell, Zenith Radio Corporation. September 12, 1955. Zenith Archives.
8. Kinney, Eugene M. (retired Senior Vice President of Zenith) and Howard Fuog (retired advertising executive of Zenith). Interview with authors, Chicago, Illinois, June 22, 1993.
9. Zenith Archives.
10. *Himalayagram* Vol. 1, No. 1 (Summer 1960). Published by World Book Encyclopedia. Contains details of the expedition and pictures of Hillary and Perkins with a model Royal 1000 Trans-Oceanic.
11. Letter from Nicholas B. Clinch, Director, 1958 American Karakoram

Expedition, to Ted Leitzell, Director of Public Relations, Zenith Radio Corporation. December 4, 1958. Zenith Archives.

12. A series of letters from John F. Stanwell-Fletcher to Ted Leitzell, Director of Public Relations, Zenith Radio Corporation. Late December 1954 through March 1955. Zenith Archives.

13. Letter from J.W. Lassiter, President, Lab Geodetics Corporation, to William S. VanSlick, Assistant Director-Government & Special Products Division, Zenith Radio Corporation. April 22, 1964. Zenith Archives.

Chapter 4: DESIGN AND STYLING OF THE TRANS-OCEANIC

1. "Design in the Chicago Mid-West." *Industrial Design Magazine* (October 1946): 56-80.
2. Holladay, Merrily Budlong. Telephone interview with author, January 19, 1994.
3. Mox, Dana. Telephone interview with author, November 1992.
4. Mox, Dana. Telephone interview with author, November 1992.
5. Memo from Gene McDonald to Steve Healey. November 9, 1951. Zenith Archives.
6. Lorenzen, Howard O. Telephone interview with author, November 24, 1992.
7. Holladay, Merrily Budlong. Telephone interview with author, January 19, 1994.
8. Mox, Dana. Telephone interview with author, November 1992.
9. *Industrial Design Magazine* (October 1956): 56-80.
10. Cascarano, Anthony J. Telephone interview with author, February 1993.
11. Kinney, Eugene M. Interview with authors, Chicago, Illinois, June 1993.
12. Cascarano, Anthony J. Telephone interview with author, February 1993.
13. Schiffer, Michael. *The Portable Radio in American Life*. Tucson: The University of Arizona Press, 1992, 221.
14. Guth, Gordon. Telephone interview with author February 1993.
15. "Portable Radios that Tune in the World," *Popular Science* (January 1967): 125-127, 206-207; and, "Shortwave FM/AM Portables," *Consumer Reports* (November 1967): 579-583.
16. Letter from Nicholas B. Clinch, Director, 1958 American Karakoram Expedition, to Ted Leitzell, Director of Public Relations, Zenith Radio Corporation. December 4, 1958. Zenith Archives.
17. Osterman, Fred J. *Shortwave Receivers: Past and Present*. Reynoldsburg, Ohio: Universal Radio Research, 1991.
18. Stender, Robert. Interview with authors, Chicago, Illinois, June 1993.
19. Stender interview, June 1993.
20. Althans, Richard. Telephone interview with author, February 1993. Althans is currently the Director of Industrial Design at Zenith.
21. Stender, Robert. Interview with authors, Chicago, Illinois, June 1993.

Chapter 5: THE END OF THE TRANS-OCEANIC

1. Bowers, William E. Interview with author, October 1993. Bowers is a former engineer with Schlumberger.
2. Hall, Luther. Telephone interview with author, May 1993. Hall is a well-known radio collector.
3. There did continue to be Trans-Oceanic product announcements and some advertising in both electronic "trade" press and in radio hobbyist-oriented press. When the pattern of advertising in general circulation magazines is examined, however, the following holds true: there were a few very small print advertisements in 1963 to introduce the Royal 3000 and then none at all until a few were run in 1968-1969, seemingly to clear the shelves of Royal 3000-1s to make way for the upcoming Royal 7000. There were only one or two small print advertisements published for the introduction of the Royal 7000 and its successor, the D7000Y. Even though the Zenith public relations firm prepared at least one major print ad for the R-7000, there is no evidence it was ever published anywhere in the general circulation press.
   This virtual absence of Trans-Oceanic advertising was true even though Zenith radios continued to be advertised in *Holiday* and *Na-*

tional Geographic. Today, it appears rather bizarre to see Zenith stereo consoles being advertised in *National Geographic* in the mid-60s in place of the finest travel portable then in existence, the Zenith Trans-Oceanic.

4. "Portable Radios that Tune the World." *Popular Science* (January 1967): 125-127, 206-207; and, "Shortwave FM/AM Portables." *Consumer Reports* (November 1967): 579-583.
5. Kinney, Eugene M. Interview with authors, Chicago, Illinois, June 1993.
6. Taylor, John I. Interview with authors, Chicago, Illinois, June 1993. John Taylor is Vice President of Public Affairs and Communications at Zenith Electronics Corporation.

Chapter 6: LINEAGE OF THE TRANS-OCEANIC

1. Schiffer, Michael. *The Portable Radio in American Life*. Tucson: The University of Arizona Press, 1992, 197.
2. Schiffer, 1991, 190.
3. Guth, Gordon. Telephone interview with author, February 1993.
4. Schiffer, 1992, erroneously reports the circumstances of the manufacture of the last portable radio manufactured in the United States. This point is quite important historically and should be corrected. Professor Schiffer states on page 224 of *The Portable Radio in American Life*:
   "One day in the Fall of 1980, although no one remembers exactly when, Zenith closed down an assembly line in Plant 2, on Kostner Street in Chicago. Zenith had moved production of the Trans-Oceanic Royal 7000 to its factory in Taiwan. Unnoticed at the time, this event ended an era. Portable radios were no longer being manufactured in the United States."
   Professor Schiffer's remarkable book is a primary resource for all radio enthusiasts. He is perhaps forgiven for, like many of us, confusing the ROYAL 7000 with the very different successor radio, the R-7000, and for citing the wrong date by almost a year. In fact, the last Zenith Trans-Oceanic manufactured in the United States, an R-7000, was produced at Zenith Plant #2, 1500 North Kostner Street, Chicago, in August 1979. Whether this date change affects the Trans-Oceanic's dubious honor as the last portable manufactured in the U.S. is still in question. Sources: Stender, Robert, and Taylor, John. Interview with authors, Chicago, Illinois, June 1993. Both Stender and Taylor are employees of the Zenith Electronics Corporation.
5. Some modern sources quote the retail price of the 7G605 Trans-Oceanic Clipper at $100. This is completely erroneous. The retail price listed in Zenith documents was $75. However, internal correspondence concerning the 7G605 in Commander McDonald's personal files indicate that a few unscrupulous distributors in the West were charging significantly more than the listed $75 for the highly sought Clipper. The Commander was very upset at those distributors and took steps to prevent future price gouging.

Chapter 7: TRANS-OCEANIC PORTRAITS

1. Kinney, Eugene and Howard Fuog. Interview with authors, Chicago, Illinois, June 1993. Kinney is a former Senior Vice President of Zenith and Fuog is a former advertising executive of Zenith.
2. Kinney, Eugene and Howard Fuog. Interview with authors, Chicago, Illinois, June 1993.
3. Kinney, Eugene and Howard Fuog. Interview with authors, Chicago, Illinois, June 1993.
4. Kinney, Eugene and Howard Fuog. Interview with authors, Chicago, Illinois, June 1993.
5. Guth, Gordan. Telephone interview with author, February 1993.
6. Cascarano, Anthony J. Telephone interview with author, February 1993.
7. Kinney, Eugene and Howard Fuog. Interview with authors, Chicago, Illinois, June 1993.
8. Stender, Robert. Interview with authors, Chicago, Illinois, June 1993. Stender is at Zenith Electronics Corporation and was involved with the development of the R-7000.
9. Althans, Richard. Telephone interview with author, June 1993.

APPENDIX I
TRANS-OCEANIC RESTORER'S TOOLS AND MATERIALS

The restoration of a Trans-Oceanic of any vintage is a rewarding project and extremely self-satisfying. The task is made easier if the restorer assembles a few items before the project begins. Although electrical problems may be far ranging and require outside help and a variety of parts, cosmetic restoration is relatively consistent among Trans-Oceanics and requires a minimum of specialized knowledge and skills. For a small cash outlay, a powerful arsenal of restoration materials can be assembled. Below is a list of materials the authors have found valuable in their own work. Sources for some of these supplies are given in Appendix II. [Specific product names are registered trade marks of the companies using them].

Cleaning Materials

Murphy's Oil Soap
Mechanic's red rags (cleaning rags)
Soft polishing clothes
Saddle soap (for leather T-Os)
000 & 0000 steel wool
Lacquer thinner

These are fairly standard cleaning materials and available almost anywhere. We have found Murphy's Oil Soap to be a noninvasive cleaner that is particularly good at removing grease and dirt. Although some restorers use 000 steel wool, we recommend 0000, since 000 can scratch some Trans-Oceanic surfaces.

Cosmetic Restoration Materials

Elmer's White Glue
Elmer's Yellow Carpenter's Glue
Super-glue
Armor All
WD-40
Kiwi Black Liquid Shoe Dye in applicator bottle
Kiwi Black Paste Shoe Wax
Kiwi Neutral Paste Shoe Wax (for leather T-Os)
Turtle Wax chrome cleaner/polish
Silver (aluminum) paint
Novus Plastic Polish # 1
Novus Plastic Polish # 2
Lacquer-Stik: Black, White and Gold
Clear acrylic plastic spray, high gloss
Brass polish
Bristle brushes
Several small, inexpensive artist's brushes
Cramolin R-5 DeOxIt contact cleaner

Kiwi products are used by museum restorers and we have found them to be especially compatible with Black Stag coverings. Novis plastic polishes are also used in museum restoration work and are readily available. Although not well known, Cramlon R-5 DeOxIt is used in high level electronics applications and we have found it to be by far the finest product available.

Useful Tools

Soldering iron and rosin core solder
Exacto-knife and blades, # 10 tapered blade recommended
Razor blades
1/4" nut driver
Side cutter pliers
Needle-nose pliers
Awl
Bench buffer/grinder
Gun blue
Plastic or nonmagnetic alignment tools

APPENDIX II
RESOURCES

Parts and Supplies

Many suppliers of materials for vintage radio restoration are not nationally advertised to the general public. The companies and individuals listed here are known to the authors and represent a good sampling of the supplies and suppliers available to you. If you work further with vintage electronics, you will discover there are many additional sources, some offering unique single services, such as knob restoration or speaker rebuilding.

Allied Electronics. 7410 Pebble Drive, Fort Worth, TX 76118. Replacement transistors.

Antique Audio (of Texas). 5555 N. Lamar, Suite H-105, Austin, TX 78751. General line of supplies.

Antique Audio (of Michigan). 41560 Schoolcraft, Plymouth, MI 48170. Capacitors, other parts.

Antique Electronic Supply. 6221 S. Maple Avenue, Tempe, AZ 85283. General line of supplies.

Antique Radio Service (Richard Foster). 12 Shawmut Avenue, Cochituate, MA 01778. Specializing in cabinet repair and restoration.

Caig Laboratories. 16744 West Bernardo Drive, San Diego, CA 92127-1904. Cramlon Contact Cleaner.

Daily Electronics. 10914 N.E. 39 Street, #B-6, Vancouver, WA 98682. Tubes.

Don Diers. 4276 N. 50 Street, Milwaukee, WI 53216. Tubes and other parts.

Electron Tube Enterprises. Box 8311, Essex, VT 05451. Tubes.

Frontier Electronics. Box 38, Lehr, ND 58460. Electrolytic capacitors sold and rebuilt.

Jackson Speaker Service. 217 Crestbrook Drive, Jackson, MI 49203. Speaker repair.

Kiwa Electronics. 612 South 14th Avenue, Yakima, WA 98902. IF filters

Mohawk Finishing Products, Inc. Route 30 North, Amsterdam, NY 12010. Refinishing supplies.

Mouser Electronics. 2401 Hwy 287 North, Mansfield, TX 76063-487. Replacement transistors.

Newark Electronics. 312-784-5100. Replacement transistors.

Old Tyme Radio Company. 2445 Lyttonsville Road, Suite 317, Silver Spring, MD 20910. General line of supplies.

Play Things of Past (Gary Schneider). 3552 W. 105 Street, Cleveland, OH 44111. General line of supplies.

Puett Electronics. P.O. Box 28572, Dallas, TX 75228. General line of supplies; manual reproductions.

Howard W. Sams & Co. 2647 Waterfront Parkway East Drive, Indianapolis, IN 46214-2041. Schematic diagrams.

Vintage TV & Radio. 3498 W. 105 Street, Cleveland, OH 44111. General line of supplies.

Club Information

Joining a radio club will put you in contact with others who enjoy vintage equipment, give you a chance to learn more about radio history and will give you access to equipment that is for sale.

This list represents many of the national and regional clubs that are available and are known to the authors. When writing for information from a club, be sure to include a self-addressed stamped envelope and tell them where you found their address.

Alabama Historical Radio Society (AHRS). 4721 Overwood Circle, Birmingham, AL 35222.

Antique Radio Club of America (ARCA). 300 Washington Trails, Washington, PA 15301.

Antique Radio Club of Illinois (ARCI). Route 3, Veterans's Road, Morton, IL 61550.

Antique Radio Collectors of Ohio. 2929 Hazelwood Avenue, Dayton, OH 45419.

Antique Wireless Association (AWA). P.O. Box E, Breesport, NY 14816.

Arizona Antique Radio Club (AARC). 8311 Via de Sereno, Scottsdale, AZ 85258.

Arkansas Antique Radio Club. P.O. Box 9769, Little Rock, AR 72219.

California Historical Radio Society (CHRS). P.O. Box 31659, San Francisco, CA 94131.

Carolina Antique Radio Society. 824 Fairwood Road, Columbia, SC 29209.

Central Jersey Antique Radio Club. 92 Joysan Terrace, Freehold, NJ 07728.

Colorado Radio Collectors (CRC). 1249 Solstice Lane, Fort Collins, CO 80525.

Connecticut Vintage Radio Collectors Club. Vintage Radio and Communications Museum of Connecticut, 665 Arch Street, New Britain, CT 06051.

Delaware Valley Historic Radio Club (DVHRC). Box 624, Lansdale, PA 19446.

Florida Antique Wireless Group (FAWG). 321 Evans Street, Orlando, FL 32807.

Greater Boston Antique Radio Collectors. 12 Shawmut Avenue, Cochituate, MA 01778.

Hawaii Chapter, Antique Radio Club of America. 95-2044 Waikalani Place C401, Mililani, HI 96789.

Houston Vintage Radio Association (HVRA). P.O. Box 31276, Houston, TX 77231-1276.

Hudson Valley Antique Radio & Phonograph Society (HARPS). P.O. Box 207, Campbell Hall, NY 10916.

Indiana Historical Radio Society (IHRS). 245 N. Oakland Avenue, Indianapolis, IN 46201.

Kentucky Chapter Antique Radio Club of America (ARCA). 1907 Lynn Lea Road, Louisville, KY 40216.

Michigan Antique Radio Club (MARC). 2590 Needmore Highway, Charlotte, MI 48813.

Mid-America Antique Radio Club. 2307 W. 131 Street, Olathe, KS 66061.

Mid-Atlantic Antique Radio Club (MAARC). P.O. Box 1362, Washington Grove, MD 20880.

Middle Tennessee Old Radio Club. Route 2, Box 127A, Smithville, TN 37166.

Mid-South Antique Radio Collectors (MSARC). 811 Maple Street, Providence, KY 42450-1857.

Mississippi Historical Radio and Broadcasting Society. 2412 C Street, Meridien, MS 39301.

Nebraska Radio Collectors Antique Radio Club. 905 W. First, North Platte, NE 69101.

New England Antique Radio Club (NEARC). P.O. Box 474, Pelham, NH 03076.

Niagara Frontier Wireless Association (NFWA). 135 Autumnwood, Cheektowaga, NY 14227.

Northland Antique Radio Club (NARC). P.O. Box 18362, Minneapolis, MN 55418.

Northwest Vintage Radio Society (NWVRS). P.O. Box 82379, Portland, OR 97282-0379.

Oklahoma Vintage Radio Collectors. P.O. Box 72-1197, OKC, OK 73172-1197.

Pittsburgh Antique Radio Society, Inc. (PARS). 407 Woodside Road, Pittsburgh, PA 15221.

Puget Sound Antique Radio Association (PSARA). P.O. Box 125, Snohomish, WA 98291-0125.

Society for Preservation of Antique Radio Knowledge (SPARK). 2673 S. Dixie Drive, Dayton, OH 45409.

Southeastern Antique Radio Society (SARS). 4830 E. Brookhaven Drive, Atlanta, GA 30319.

Southern California Antique Radio Society (SCARS). 656 Gravilla Place, La Jolla, CA 92037.

The Southern Vintage Wireless Association (SVWA). 1005 Fieldstone Court, HSV, AL 35803.

Vintage Radio and Phonograph Society (VRPS). P.O. Box 165345, Irving, TX 75016.

Western Wisconsin Antique Radio Collectors Club (WWARCC). Route 1, Box 182-A4, Stoddard, WI 54648.

West Virginia Chapter, Antique Radio Club of America. 405 8th Avenue, St. Albans, WV 25177.

Vintage Radio Hobby Buy/Sell Publications

The periodicals listed here are the leading national sources of vintage equipment for sale or trade. Sample copies are available for several dollars; a post card will bring subscription information.

Antique Radio Classified. P.O. Box 2, Carlisle, MA 01741.

Ham Trader Yellow Sheets. P.O. Box 15142, Seattle, WA 98115.

Wireless Trader. 4290 Bells Ferry Road, Suite 106-36A, Kennesaw, GA 30144.

Transistor Network. RR1, Box 36, Bradford, NH 03221.

Electric Radio. P.O. Box 57, Hesperus, CO 81326.

Preface to the Second Edition

In the years since the original publication of this book, we have been able to spend a great deal more time in the Zenith Archives and with other primary sources. We have also closely followed the activities of the community of Trans-Oceanic enthusiasts. Throughout our efforts on four additional Zenith related books, we have kept in mind the notion of a second edition of our first love, the Trans-Oceanic book.

As we began our efforts on this Second Edition by reviewing the Trans-Oceanic materials in the Zenith Archives, we were rather surprised at how thoroughly we had covered the history of the Trans-Oceanic in our original effort. Indeed, we have found only one important addition to that history that we can document from primary sources: the story of The Quiz Kids radio program and the company president, Eugene F. McDonald's response to the producer of the Quiz Kids on the early development of the Trans-Oceanic. Probably one of the most important discoveries over the past decade has been the identification of two new Trans-Oceanic models, the Royal R-520A/URR and the D7000Y. And we feel it is time that we document the Trans-Oceanic competitors that attempted to steal the market from Zenith.

We have also presented several additional restoration tips and strategies that should prove useful, and we have updated the Price Guide to Trans-Oceanics and added a rarity category and production figures. We have also provided our "Flea-Market Guide to Zenith Trans-Oceanics," and close with some observations about the many internet websites dedicated to these wonderful radios. We want to give special tanks to Martin Blankinship for assistance in collecting some of the material in this edition.

An Important Historical Footnote

The Quiz Kids and the Trans-Oceanic

The Quiz Kids was a very popular radio program during the Golden Age of Radio. The show first aired on the NBC Red Network on June 28, 1940, as a summer replacement for blind pianist Aleck Templeton. The initial series was ten weeks long and was heard over 33 stations. The program was renewed for the regular season in September 1940 and 12 more stations were added to the network. Zenith supplied a number of 6G601 Universal portable radios to Wade Advertising Agency, agents for Miles Laboratories, makers of Alka-Seltzer and the sponsor of The Quiz Kids program. Miles gave away a Zenith portable if a listener's question was used on the program. During the program, Master of Ceremonies Joe Kelly asked the panelists 10-12 listener questions, meaning that in a 39-week season, roughly 400 Universals were given away each year. Later in the show's run, if the listener's question actually stumped the kids, the listener was awarded a large Zenith, such as a radio-phonograph console.

In April 1941, The Quiz Kids went on the road to originate from Hollywood, appearing for two weeks on the Jack Benny Show. Zenith presented two 6G601M Universals to Jack Benny and Mary Livingston, on the air, during the April 16, 1941 show. By that time, ratings for The Quiz Kids had reached the point that a line of "Quiz Kids" merchandise was possible. Among the ideas proposed for merchandise was a low-cost, Zenith-produced, "Quiz Kids" radio — but Zenith did not act on the idea.

The Quiz Kids began giving away the 7G605 Trans-Oceanic Clipper on the January 14, 1942 show, rather than the Zenith Universal. According to a January 20 memo from the Wade Advertising Agency to Zenith, 25,000 letters containing potential questions were received in response to the Trans-Oceanic offer, the greatest number of letters the show had ever received.

President McDonald responded to the Wade Agency letter on January 26, 1942:

Our advertising department has shared with me your letter of January 20.

I am delighted to see the pulling power of this Trans-Oceanic Portable on your program.

You might be interested in some facts in connection with this radio which I am giving you below:

I authorized the production of this radio two years ago the second day of last August. Since then the laboratory has submitted to me over twenty models, all of which I rejected, until this last present model was produced.

When the retail price had to be placed at $75, we found our distributors didn't believe the product would sell – our dealers didn't believe it – no one believed it, except the public. As a sample of the lack of interest in this radio, our New York distributor placed an initial order for 300. A week after he received his first shipment and saw the way it moved off the dealers' shelves, he raised his order to 1,700 and last week telegraphed in another order for 3,000. In other words, with an initial order of 300, he is now up to 4,700 and begging for deliveries.

It is one of the most phenomenal jobs we have ever turned out from the standpoint of performance and the beautiful part of it is that we have no competition as there is nothing like it on the market.

I am enclosing herewith a copy of the publicity that went out to the daily newspapers only a week ago today.

I appreciate the splendid job you are doing for us.

I want to record my thanks.

Sincerely,
Eugene F. McDonald, Jr.

Just before civilian radio production ceased in April 1942, Zenith provided a year's supply of 7G605 portables to the radio program that were kept in a separate warehouse from those in the Zenith vault. The Quiz Kids supply eventually ran out, and starting with the June 27, 1943 program Zenith began issuing a gold certificate redeemable for one post-war Zenith Trans-Oceanic Clipper.

Trans-Oceanics, Companions, Cousins, and Competitors

New Models

When the first edition was published, there were rumors in the radio community of a "second" military morale radio, in addition to R-520/URR. Several years ago, we were stunned to find a mint example of the second military model, the R-520A/URR in the bowels of Zenith's headquarters in Glenview, Illinois. This particular radio had apparently never left the plant, never had the military ID tag attached and is the radio featured in our photos here. Recently, we have also seen several R-520A/URR's preserved in the enthusiast community.

In the first edition, we classed the D7000Y as a variant in the Royal 7000 family of radios. It looked like a Royal 7000 and functioned much the same, so our decision was understandable. After a good deal of thought and consultation with the former Zenith staff, we have determined that the D7000Y was, clearly, a separate model. It had a new name (no "Royal" here!) and was provided with a completely new chassis (500MDR-70 vs. the old 18ZT40). The D7000Y clearly deserved designation as a separate model.

Companions and Cousins

The Global 6G004Y was mentioned in our first edition in the discussion of the 8G005Y Post-War Trans-Oceanic. We discovered the unique history of the Global in our later research in the Zenith Archives. As we have discussed in our other books, Commander McDonald exercised control over virtually all aspects of Zenith product development, manufacturing and marketing; this was clearly demonstrated with the production history of the 6G004-Y. This portable was produced in early 1947, apparently at the behest of the sales department and with only reluctant approval from McDonald who was concerned about the competition of this inexpensive radio with the Trans-Oceanic. As Global production began, specialized parts had been acquired for an initial run of 15,000 units; at the end of that run, McDonald personally halted Global production permanently, making the Global one of the rarest Zenith portables.

The Meridian L507 was a second attempt by the Zenith sales staff to tap what they thought was a market for medium-priced shortwave portables. Despite being similar in size to the L600 Trans-Oceanic that it resembles, and despite providing two shortwave bands, the set neither performed nor sold well. McDonald likely reminded the sales department of the history of the Global six years earlier and took similar action with the Meridian.

The only evidence in the archives of the Zenith Receiving System ZX-5 are the images reproduced here. The ZX-5 appears to be a repackaged Royal 1000/3000, though the Zenith narrative claims coverage up to 200 megahertz. The narrative also mentions "two of them may be carried on the standard marine pack board." We believe that this refers to a standard U.S. Marine Corps pack board. The ZX-5 was very likely a mock-up to support either an unsolicited proposal to or an informal request by the U.S. Navy and Marines. As far as can be determined, the ZX-5 was stillborn.

The little-known 1965 Zenith Explorer M660 was introduced as a headliner in the 1965 Zenith Line; it was a newly designed all-wave tabletop radio similar to the Hallicrafters S-120 or the National NC-60. It is difficult to imagine why post-McDonald Zenith assigned engineering talent and production capacity to a new tube circuit, seven long years after the first transistorized radios joined the Zenith Line. The Explorer emulated its more prestigious cousin, the Royal 3000 Trans-Oceanic in most of its aesthetic features. Relatively few were produced.

While looking through radio catalogs at Zenith headquarters, the authors were shocked to discover this distant cousin to the Royal 1000 Trans-Oceanic of the late 1950's. In Zenith history, the Model 1000 designation had been sacred, used only for the legendary Stratosphere console of the 1930's and for the equally legendary Royal 1000, the first transistorized T-O of the 1958 Line. The 1977 Royal G1000Y was a beautifully designed radio in the tradition of all of the solid state Trans-Oceanics: rendered in brushed chrome, black leather, and black plastic. The frequency coverage though, completely ignores the shortwave bands of the T-O's and concentrates on VHF and UHF. The authors had never seen or heard of an actual Royal G1000Y until the radio shown on these pages appeared on eBay, listed as a Trans-Oceanic.

Competitors

The Trans-Oceanics had few competitors during the tube years. This was likely due to the difficulty of getting battery-powered tubes to perform at shortwave frequencies. In the last few years before transistorized portables arrived in the market (1952-58) two main rival manufacturers enter the multi-band portable marketplace: Hallicrafters and the Zenith nemesis, RCA. In the mid-1950's Hallicrafters began what proved to be an ill-conceived move into the consumer electronics market. One of Hallicrafters major offerings was a series of "Trans-World or "World-Wide" suitcase portables very like the T-O's of that period. The TW-500and 600 were direct competitors to the Zenith L507 and the TW-1000, 1000A and 2000 were direct competitors for the 600 series T-O's. Probably it was the lack of a retail dealer network like Zenith's and RCA's that doomed the Hallicrafters' efforts.

For many years, the RCA Victor all-band portable Strato-World was a feature of the RCA Line. The original Strato-World, the 1954 Model 3-BX-671, was reportedly designed at the personal request of RCA President David Sarnoff to drive Commander McDonald's personal radio, the Trans-Oceanic, completely from the market. A grand example of late Art Deco industrial design, the 1954 RCA Victor Strato-World was arguably on an equal footing with the elegant leather-clad Trans-Oceanics of the 600 Series. However, even the giant RCA was no match for the fabled T-O and down through the years each succeeding model of Strato-World fell farther from the high standards of the Zenith Trans-Oceanic.

For a variety of reasons, the advent of transistorized portables in the late 1950"s also marked the beginning of an explosion in competitors for the Trans-Oceanic. Soon there were literally dozens of competitors, not only from rival American manufacturers, but also from leading European marques and a myriad of manufacturers in the Far East. Many of these radios were inexpensive, performed poorly and were no real competition to the T-O. However, top-of-the-line sets from major foreign manufacturers such as Grundig, Nordmende, Panasonic and Sony were excellent, with performance that rivaled and later exceeded the transistorized Trans-Oceanics.

Trans-Oceanic R-520A/URR. Courtesy of Zenith

Trans-Oceanic R-520A/URR: New Model

The R-520 was the militarized version of the H-500 model Trans-Oceanic; this radio, the R-520A was a slightly militarized version of the 600 Series Trans-Oceanics. There were 2,973 R-520A/URR radios produced in 1956, making this model by far the rarest Trans-Oceanic. The R-520A was based on the L or R600 model, with very few modifications (no dial light, headphone jack accessible from the rear only.) The cabinet was covered with an early vinyl cloth in Army green and "USA" was stamped on at least one end of the cabinet in large letters; the main plastic face plate was only slightly modified from the civilian 600 Series unit.

The earlier military model, the R-520, was a very militarized/ruggedized radio: a spare set of tubes and tuning tools were clipped inside the cabinet; there was special shielding added; special combined batteries were used; the civilian paper capacitors were changed to more rugged plastic units; mildew and damp proofing was taken seriously; special military sealed unit packaging was developed. The R-520A provided none of these "mil-spec" upgrades to the civilian 600 model upon which it was based. Normal civilian batteries were used and, except for the minor modifications mentioned, the chassis was an early 600 Series radio. L600 or R600 service data may be used for the R-520A with confidence.

Note that at least three R-520A/URR radios exist which never carried the military aluminum identifier tag (with black fill.) The unit featured on these pages was still the property of Zenith when it was photographed in the late 1990s. Midwestern hobbyists hold at least two other untagged units. Other 520As, presumably initially owned by the US Army, did have standard military tags attached beneath the latch of the front cabinet door.

Trans-Oceanic R-520A/URR.

Trans-Oceanic R-520A/URR.

Trans-Oceanic D7000Y. *Courtesy of John H. Bryant.*

Trans-Oceanic D70000Y: New Model

The D7000Y was introduced in December 1972 for the 1973 Line and remained as Zenith's premiere portable radio until the R-7000 was introduced in the 1979 Line. The D7000Y introduced an updated chassis, the 500MDR70. This chassis shared some transistors types with the old 18ZT40Z chassis of the Royal 7000 and some were updated designs. From a user's point of view, the primary difference between the Royal 7000 and the new D7000Y was the addition of a tunable Weather Band. Enthusiasts readily identify the D7000Y model by its dramatic black compass plate which lays just below the silver speaker grille when the radio is in use; the compass plate on the Royal 7000 had been gray.

By the mid 1970s, the mass-circulation advertising budget for the radio division of Zenith must have been quite small, as advertising for the flagship Trans-Oceanic was miniscule. A single campaign was structured for the D7000Y that featured the phrase "'Round the World Tour." Two print ads were designed and a few were placed in magazines. The two ads are reproduced on pages 42 and 113 of this book. Retail price for the D7000Y was between $280 and $300. Please refer to pages 110 through 117 for further details of the Royal 7000, Royal 7000-1 and the D7000Y. Note that the proper model title for this receiver does not contain the word "Royal."

Sadly, the D7000Y was the last Trans-Oceanic that provided the Commander's dedicated bandspread dials for listening to international shortwave broadcasting. This capability, precious to the user but unappreciated by most radio design engineers, makes the D7000Y most certainly the best Trans-Oceanic ever produced.

Trans-Oceanic D7000Y.

Trans-Oceanic D7000Y.

Trans-Oceanic D7000Y.

Global 6G004Y. *Courtesy of John H. Bryant*

Global 6G004Y.

The Global 6G004Y: Cousin

Like the Trans-Oceanic 8G005Y and its Companion, the broadcast band-only Universal 6G001Y, the Global was designed during World War II for the immediate Post-War Line. The cabinets and face-plates for all three sets were designed by Robert Davol Budlong and approved by the Zenith leadership in 1944. With only 15,000 produced in a single production run, the Global is exceedingly rare today.

The Global shared most of its circuit design with its near-twin, the Universal 6G001Y and provided an identical tube complement of six-Loctal tubes (1LN5 R.F., 1LA6 Converter, 1LN5 I.F., 1LH4 Det.-Amp., 3Q5GT Power Amp., 117Z6G/GT Rectifier). The RF section was unsophisticated and the Global tuned from 535 to 1620 kc and 9400 to 12000 kc. Service information and the circuit diagrams for both the Global, chassis 6C41 and the Universal, chassis 6C40 are found in Zenith Service Manual, Volume 3, pages 72-75. The Global was advertised only briefly at the introduction of the Post-War Line: Saturday Evening Post, May 5, 1946, p. 121; Better Homes and Gardens, May 1946, p.13.

Global 6G004Y.

Global 6G004Y.

Meridian L507. *Courtesy of John H. Bryant.*

Meridian L507.

The Meridian L507: Cousin

The Meridian was produced for the 1954 Line in what appears to have been very limited numbers. Like the Global, it was designed for the multi-band portable market segment lower in price than the Trans-Oceanic. Although the Meridian provided an identical tube compliment to the vaunted L600 Trans-Oceanic, to meet its pricing goal it was designed with a low parts count and a rudimentary RF section. Those decisions along with an unsophisticated non-bandspread dial, doomed the L507 to an early death. It tunewwwbroadcast band and 2 to 18 MC in two shortwave bands. The Meridian L507 is quite rare today and of historical interest to some Trans-Oceanic enthusiasts. The chassis is identified as 5L42 and the circuit diagram and alignment information is found in Zenith Service Manual, Volume 5, p.19. There was a limited amount of print advertising of the Meridian in late 1953 and early 1954, primarily in mass-circulation magazines.

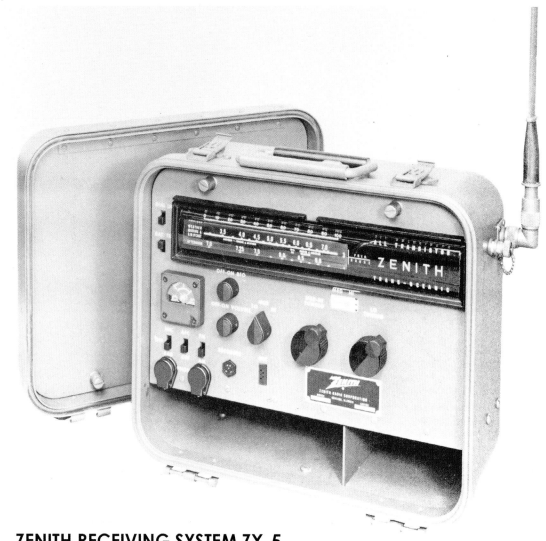

ZENITH RECEIVING SYSTEM ZX-5

Zenith Receiving System ZX-5

Zenith Receiving System ZX-5.

ZENITH RECEIVING SYSTEM ZX-5 PROVIDES radio reception over a range of 0.54 to 200 Mc, in 10 bands. The system includes a tape recorder capable of recording up to three hours of information: the operator may monitor and record radio signals, or his own voice. A beat frequency oscillator provides for monitoring and recording CW signals. Relative signal strengths can be compared, and the system serves as a radio direction finder when used with a direction-finder loop. All controls and indicators are designed for simplicity and easy access. Storage space is provided in the lid for spare tape magazines, spare batteries, and other accessories.

The system is packaged in a fiber glass case with dimensions of approximately 10-1/2"H X 13"W X 9"D, including cover. The system is light enough so that two of them can be carried on a standard marine pack-board.

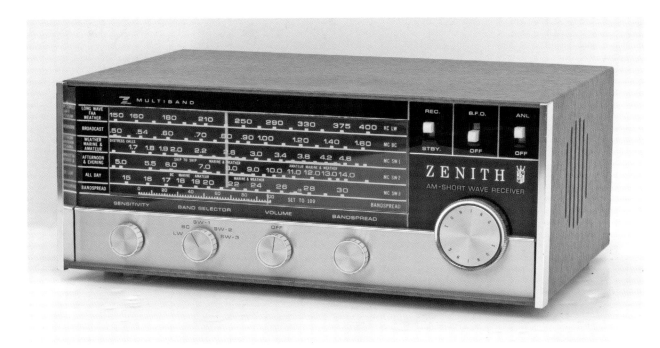

Explorer M660. *Courtesy of John H. Bryant.*

The Explorer M660: Cousin

The new Explorer M660 headlined the sales brochures for the 1965 Zenith Radio Line. It offered 5 bands LW, BC, 3 SW and was built around a feature-filled chassis supporting five tubes, one diode and a selenium rectifier. The Explorer measures 6"H, 14 ¾"W and 9"D. Unfortunat ely, the tube technology of the new Explorer had been obsolete in the marketplace for at least five years. Further, each shortwave band covered about 10 MC, making tuning and retuning a specific station very difficult. The Explorer also appeared in the 1966 Line. The Explorer shared knob design and all front panel graphics, including the basic design of the dial, with the Royal 3000 Trans-Oceanic. The Explorer may have been a one production run product which was not at all successful in the marketplace. It is very rare today.

Royal G1000Y.
Courtesy of John H. Bryant.

Royal G1000Y: Cousin

Other than its companion Trans-Oceanic D7000Y, the Royal G1000Y was the top portable radio in the 1977 and 1978 Zenith Lines. The Royal G1000y was a 6-band portable, featuring FM/AM/PSB-Lo/PSB-Hi/PSB-UHF/VHF-Air and was powered by D-cell batteries or AC, with a wall-mounted transformer. It measures approximately 9" x 12" x 3.5" Note that the front of the case carries the title "Royal 1000." It sold for $119.95 in 1977, is extremely rare and may have been produced in only a single production run of a few thousand units.

RCA Strato-World.

RCA Strato-World 3-BX-671: Competitor

The first Strato-World, introduced in the mid-50s, was purportedly designed to sweep the Trans-Oceanic from the market place. With four dedicated bandspread dials covering the international broadcast bands, and an elegant leather case in late Art Deco style, it was probably the only competitor of the tube years to be at least the equal of the Trans-Oceanic. It is interesting to note that the 3-BX-671 offered a tube complement, including the now-rare 1L6, that is identical to that of most of the Trans-Oceanics.

RCA Strato-World.

RCA Strato-World. *Courtesy of John H. Bryant.*

RCA Strato-World.

Hallicrafters World-Wide TW-1000.

Hallicrafters World-Wide TW-1000. *Courtesy of John H. Bryant.*

Hallicrafters World-Wide TW-1000: Competitor

The Hallicrafters World-Wide may have been a better radio that its competitor Trans-Oceanic in the first two years of its 1952-56 production. Equipped with a better dial, an innovative tuning turret and a similar 5-tube circuit, the World-Wide was still no match for the mystique and marketing behind its powerful Trans-Oceanic competitor.

Hallicrafters World-Wide TW-1000.

Royal 1000. *Courtesy of Howard Fuog.*

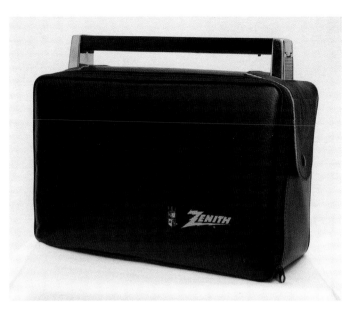

Royal 1000.

Royal 3000. *Courtesy of Zenith.*

Royal 3000.

ACCESSORIES

Over the forty-year life of the Trans-Oceanic radio line, Zenith produced relatively few accessories and the first edition of this book covered most of them quite adequately. During the solid-state years though, there were several accessories produced by third parties, specifically for the Trans-Oceanics. These accessory products were generally developed under the guidance of Zenith and were marketed by their manufacturers directly to major Zenith wholesale distributors and retail dealers. Clear vinyl dust covers, cut to fit specific Trans-Oceanic models certainly were available for the Royal 7000 Series and probably for the Royal 1000 and 3000 radios, as well.

The most interesting third party accessory, available for the Royal 1000s and the Royal 3000s, at least, were specially manufactured, fitted padded carrying cases of top-quality black vinyl. These cases were designed to provide maximum protection for these very expensive radios and were configured so that the radio could be used while swaddled in the carrying case. One of the two elegant cases shown here is courtesy of Mr. Howard Fuog, one of the legends of Zenith during the Post-War years.

Trans-Oceanic 8G005Y Design Prototype: 1944. *Courtesy of Zenith.*

Universal 6G001Y Design Prototype: 1944. *Courtesy of Zenith.*

Industrial Design Prototyping: 1944

During the Golden Years of the latter 1930s, each succeeding annual Line of Zenith radios was developed in a year-long routine of structured meetings between sales and engineering executives along with Commander McDonald, Chief Engineer Gustafson and Vice President Robertson. Most of these meetings were also attended by long-time Zenith industrial design consultant Robert Davol Budlong and possibly several others. These meetings began almost two years in advance of the Introduction of the new Line to the public. About a year in advance of Introduction, the industrial design consultant (Budlong on all but the wooden cabinets of some models) would stage a dramatic presentation of full-scale mock-ups or prototypes of the proposed new Line.

With the advent of WWII, civilian production of the 1942 Zenith Line was halted in April of 1942 and the plant went to full war production. Very soon, however, a few Zenith engineers whose lack of security clearance prohibited them from involvement with "war work" were set to planning the Post-War Line. This team also included industrial designer Budlong, whose very poor health precluded military service, and whose skills were not really relevant to war work. When Commander McDonald later discussed those days, he proudly pointed out that not one single hour of 12 hour per day war work went undone to accomplish this unique post-war planning effort. In 1944, Robert Davol Budlong presented his cabinet design prototypes for what became the 1946-47 Zenith Post-War Line. The three photos above are among the few remaining in the Zenith archives that record this design and prototyping process.

The design of the front panels and cabinets of the Universal and the Global were produced identically to their original prototypes. The cabinet of the production Trans-Oceanic was slightly modified from its prototype, probably in response to budgetary or production concerns.

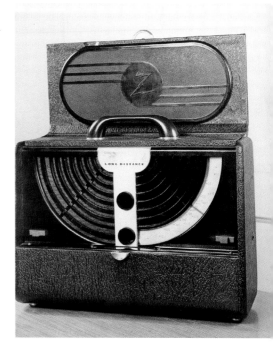

Global 6G004Y Design Prototype: 1944. *Courtesy of Zenith.*

Restoration Notes

The 1L6 Tube

Unfortunately, tubes fail, and for the most part, replacements have been fairly easy to find. With the coming of the age of transistors, tube manufacturers switched to other products and replacement tubes started to become more difficult to acquire. Some tubes, such as the major power tubes for older high fidelity equipment, disappeared into the massive Japanese collector frenzy for tube-based audio systems of the early 1990s. One would think that the small, low-power tubes of the Trans-Oceanic would be relatively immune from shortages, but such is not the case. Trans-Oceanics, from G500 through the 600 series and most of their tube-era competitors, used a pentagrid converter, the 1L6, that was not used in a large number of other radio models; it therefore was never made in great volume. As a result, as tube radios faded from the market, the supply of 1L6s also faded. In fact, they were already scarce when this book was first published and they are presently almost impossible to find--$50 is a bargain, if you can find one. What to do? Several alternatives do exist, developed by T-O fanatics with technical ability.

Direct Replacement. A 1U6 is almost a direct replacement for the 1L6 except for its filament resistance. In order to make this direct swap work, a 56 ohm resistor has to be placed in the filament string. A quick scan of a tube catalogue, however, reveals that the 1U6 is just as difficult to obtain and just as costly as the 1L6. The 1U6 does turn up more often at radio flea markets than the 1L6—everybody seems to know about 1L6s.

Other experimenters have directly replaced the 1L6 with the much more plentiful 1R5. There is a loss in sensitivity in this replacement, especially at the higher frequencies, that can partially be offset with a realignment of the receiver. If your goal is to just enjoy listening to your Trans-Oceanic, rather than DXing with it, the 1R5 replacement will be a quick and easy fix for you.

Several experimenters have used the loctal based 1LA6 by developing a loctal socket to miniature socket adaptor. Several sites on the web will guide you through this replacement and at least one site sells ready-made units. Reports indicate that performance is equal to the 1L6, but a new alignment is suggested for peak sensitivity.

Solid State Replacement. Plans for building solid state replacements on a removable miniature tube base are available from various hobby sources, as are ready-made units. The appearance of the replacement unit, as you might expect, is not the same as that of the 1L6. Some T-O enthusiasts use the solid-state replacement for casual listening and replace the 1L6 for display of the radio. At least one well regarded supplier manufactures a full set of solid state tube replacements for the Trans-Oceanic.

Black Stag

Black Stag was Zenith's official name for the vinyl/cloth covering for most of the tube Trans-Oceanics. As many older and worn tube-era models enter the market, Black Stag restoration has become a much discussed topic among T-O enthusiasts. Although several manufactures and dealers have carried very similar material in the years since our book was first published, we have yet to encounter an exact Black Stag match in texture and thickness. Replacing the Black Stag is not a job for amateurs—it is almost impossible to totally remove all the old material and glue from the wood case and closely fitting the new material is an art form itself. Instead of replacement, the authors suggest that repairs be made to the existing material if at all possible. Although we have been criticized for suggesting the use of black shoe dye and polish on the Black Stag, it was first suggested by Zenith and it does do wonders for old cases; as with any restoration project, you have to determine how "purist" you want to be. Some enthusiasts would rather retain the old covering as authentic, while others want it to look new. We also note that a number of Trans-Oceanic enthusiasts have had success with mink oil applications to moderately worn Black Stag.

Date Stamps on Speakers

Not long after the T-O book was first released, word began to circulate in hobby circles that Trans-Oceanics could be dated by referencing the date stamp on the speaker. We have searched many boxes of factory records and interviewed a number of Zenith employees and still have no definitive answer as to whether the date stamp on the speaker documents the date of the manufacture of the speaker or that of the finished radio.

When the legendary Hugh Robertson joined Zenith in December 1924, his first task was to get the purchasing and distribution system in check. He found Zenith produced radios for wholesale distributors without any idea of how many they would buy or could take. This often left Zenith unable to fill orders for a specific model or with a large and expensive inventory of parts and finished products to manage. Robertson started a program which required the distributors to provide a commitment for ordering specific numbers of products and required them to update that commitment quarterly. Once these figures were known, parts ordering at Zenith could be regulated to match known commitments, and if all procedures were followed, by year end, all commitments would have been met and the factory would be empty. Determining advance production and sales figures—an inventory control program—seems very basic now, but the concept was new in the radio industry when Robertson introduced it. It quickly spread, becoming an industry standard. His procedures, which took several years to mature, also allowed Zenith to increase advertising for slow-selling models, regulate the size of the factory work force, and minimize capital invested in outdated parts. The financial stability that resulted from the Robertson system built a fiercely loyal group of distributors and dealers and was a major component in Zenith's ultimate success. Zenith used Robertson's system through its lifetime; it would therefore seem logical that since components did not stay in inventory very long at the factory, a date stamp on a component such as a speaker would be a good indication of the age of the radio bearing it—if not the exact date, then a close date, unless, of course, the component has been replaced at some point in the life of the radio.

The authors copyrighted this 'Flea Market Guide to Zenith Trans-Oceanics" in 1995 and handed it out at radio meets and flea markets as early promotional material for this book. One of the authors is a biologist and uses a key of this style to "key out" or identify unknown organisms. We have been frequently asked for copies of this key, so we thought it would fit nicely with this second edition.

The Trans-Oceanic on the Web

There are presently a dozen or so websites dealing with topics related to Trans-Oceanics. Some are technical, some are chatty and some are photo galleries, but all are helpful to at least some degree. Simply type "Zenith Trans-Oceanic" into your search engine and away you go. But, a word of caution: as with websites dealing with almost anything, several of the websites contain the thoughts of self-professed "experts" who seem to consider themselves the fount of all Trans-Oceanic knowledge. Over the past 15 years we believe we have researched all the original memos, work orders and factory order information that still exists, and other than an occasional memo that surfaces from an obscure source, we believe we have read all the known official documents relating to the history of the Trans-Oceanic. We have found it necessary to be cautious of the "new" history disclosed on some of these sites. We also resent the fact that some sites use photographs scanned from this book and repeat information only found here as their own. There is no way, in many cases, that this information could be known by anyone, unless they had access to official Zenith historical materials, which, as of this date, they have not. The danger is that this informal, inaccurate "history" could become officially accepted in the scholarly community as fact. We have seen a number of absolutely incorrect historical statements presented with such authority that unsuspecting readers might accept them as correct. The number of Trans-Oceanics produced is a frequent topic, for example. We see wide variations in these numbers presented on web sites, especially in the case of the 7G605, with no explanation as to where these numbers came from, but with such authority they appear to be valid to the unsuspecting reader. We have held the actual, original production reports in our hands and are very confident with the numbers we have presented in this book.

Where we have found these internet resources to be most useful is with technical information. A number of very good Trans-Oceanic technicians have appeared on the internet who provide excellent technical advice and suggest fixes for failed components. Large spaces on some sites deal with everything from replacing the Black Stag to restoring whole radios. Some of these sites also provide copies of schematic diagrams and owner's manuals.

The Web discussion sites also yield a great deal of colloquial information that can be verified through the large number of site participants. For example, some time ago, someone noticed that 7G605s with the Sailboat grill have "Special" on the bottom of the dial, whereas the Bomber models do not. The authors have seen no hard evidence to support this in any of the Zenith files, but from the examination of collector's radios and the responses of participants of the web sites, this appears to be a valid discovery.

Since web addresses change quite often, we have not provided them here; they are easily obtained with a web search. An additional excellent information source is the Trans-Oceanic CD available from Radio Era Archives. This $89 CD provides manuals, schematics and service data for all T-Os in one convenient and compact location.

PRICE GUIDE

Model	Type	Production	Years Produced	Rarity	Value
7G605	Tube	34,655	1942	D	4-5
8G005Y	Tube	110,567	1948-1949	B	3
G500	Tube	89,681	1950-1951	C	3
H500	Tube	245,544	1951-1953	B	2-3
R520/URR	Tube	27,218	1952-1953	D	4-5
600 Series	Tube	269,099 *	1954-1963	C	2-3
R520A/URR	Tube	2,973	1956	D	5-6
Royal 1000/1000D	Trans.	165,721 **	1958-1963	B	3-4
Royal 3000	Trans.	Unknown	1963-1971	A	2-3
Royal 7000-Y	Trans.	Unknown	1969-1970	C	3-5
D 7000-Y	Trans.	Unknown	1973-1978	B	3-5
R 7000	Trans.	75,000	1979-1981	C	4-5

* Of this total, 47,018 were leather covered
**thru May 31, 1965

Production: The only production figures that still exist at Zenith for the Trans-Oceanic are figures hand tallied from official records prior to July 1965 in an effort to find and celebrate the fifty millionth Zenith product. The data sheets indicate that 1,029,744 Trans-Oceanics had been produced by May 31, 1965. There are no further known production records in existence and we will probably never known how many transistor Trans-Oceanics were produced.

Rarity
A	Very Common
B	Common
C	Somewhat uncommon
D	Uncommon to Rare

Rarity: Based on long range study of Trans-Oceanics that appear at vintage radio shows, eBay, and antique stores.

Value
1	Less than $80
2	$80-120
3	$120-180
4	$180-270
5	$270-400
6	$400-600

Value: Based on average resale values within a category. Condition is of prime consideration in determining price.